Hassane Kamaria
Roger Edmond

**Mangrove ecology and carbon sequestration assessment:
Comoros**

Hassane Kamaria
Roger Edmond

Mangrove ecology and carbon sequestration assessment: Comoros

ScienciaScripts

Imprint
Any brand names and product names mentioned in this book are subject to trademark, brand or patent protection and are trademarks or registered trademarks of their respective holders. The use of brand names, product names, common names, trade names, product descriptions etc. even without a particular marking in this work is in no way to be construed to mean that such names may be regarded as unrestricted in respect of trademark and brand protection legislation and could thus be used by anyone.

Cover image: www.ingimage.com

This book is a translation from the original published under ISBN 978-613-8-45883-8.

Publisher:
Sciencia Scripts
is a trademark of
Dodo Books Indian Ocean Ltd., member of the OmniScriptum S.R.L Publishing group
str. A.Russo 15, of. 61, Chisinau-2068, Republic of Moldova Europe
Printed at: see last page
ISBN: 978-620-4-07212-8

SUMMARY

INTRODUCTION	6
PART 1	8
PART 2	18
PART 3	27
PART 4	56
CONCLUSION	65
BIBLIOGRAPHY	67
ANNEXES	73

ACKNOWLEDGEMENTS

We would like to thank Allah the Almighty who granted us life and gave us the chance and the courage to write this thesis.

This work is one of the results of the collaboration between the Faculty of Science and Technology of the University of Comoros, the Department of Plant Biology and Ecology and the Department of Animal Biology of the Faculty of Science of the University of Antananarivo.

This thesis could only be undertaken and completed thanks to the support and help of many individuals and companies who we are very pleased to thank here.

Firstly, to the Department of Animal Biology and the Department of Plant Biology and Ecology which initiated this DESS- SE training.

-   To all the members of the jury, especially :

To **Professor Noromalala RASOAMAMPIONONA RAMINOSOA,** Professor at the Faculty of Sciences of the University of Antananarivo and Teacher at the Department of Animal Biology, for having brought criticisms and remarks in order to improve our work, and for having done the honour of chairing this jury.

To **Doctor Edmond ROGER**, Lecturer at the Faculty of Sciences of the University of Antananarivo, who despite the distance and time, gave us his advice and suggestions and ensured our supervision. Without his help, this work would not have reached its end. Please accept our deepest gratitude.

To **Dr. Emilienne RAZAFIMAHATRATRA,** Senior Lecturer at the Faculty of Sciences of the University of Antananarivo, and Lecturer at the Department of Animal Biology, for agreeing to examine this work.

-   To **Doctor Daniel RAKOTONDRAVONY**, Coordinator of the DESS-SE training
-   To all the teachers of the DESS-SE Conservation Biology.

Then, to the persons in charge of the Herbarium of Comoros who allowed us to carry out our research within their laboratory and who reserved for us, during our training course, an unforgettable welcome and sympathy that we will not deprive ourselves of mentioning.

Our thanks also go to :

> Mrs. Andilyat M Abderemane, my professional supervisor, who despite his time, provided technical advice and constructive comments during our field trips and at the time of writing.

> Mr. Ahamada Daroussi and Ms. Ramadhoini Ali Islam who provided constructive advice during the drafting process.

> Mr. Nourdine Mirhani for his help and technical advice in the field.

We would also like to pay tribute to and express our gratitude to :

My colleagues and friends, Soulé Ahamada Ibrahim, Ben Anthoy Bacar, Asma-Ilhousna Mohamed, El-yamine A Mohamed, Mmadi Soufiane, Said Mmadi, Ahmed M Nadjim,

Bahagat S Abdoulkarim, who helped us efficiently in the field and in writing. May they know how much we appreciate their cooperation.

This brief is particularly dedicated to :

- My parents, Zainaba Msaidie, my father-in-law Ali Ahmed and my father the late Hassane Mdoihoma who showed us the way to success.
- My big brother Mahamoud Hassane who gave me the courage, the support and the financial means to follow this training.
- My great sister Harmia Hassane and my brother-in-law Mihidhoir Said Bacar who never stopped supporting me during my university studies until the elaboration of this thesis.

To all those who, from near or far, have contributed to the realization of this work.

Thank you very much!

ACRONYMS AND ABBREVIATIONS

ACA	: Action Comores Anjouan
AIDE	: Association d'Intervention pour le Développement de l'Environnement
ANACM	: Agence Nationale de l'Aviation Civile et de la Météorologie
CNDRS	: Centre National de Documentation et de Recherche Scientifique
COI	: Commission de l'Océan Indien
CO2	: Dioxyde de carbone
COSeP	: Centre des Opérations de Secours et de la Protection civile
DESS-SE	: Diplôme d'Etude Supérieure Spécialisée en Science de l'Environnement
DGE	: Direction Générale de l'Environnement
DGIC	: Direction Générale de la Coopération Internationale
DNP	: Direction Nationale du plan
DHP	: Diamètre à Hauteur de Poitrine
FAO	: Food and Agriculture Organization
GES	: Gaz à Effet de Serre
GEF	: Global Environment Facility
GPS	: Global Position System
HDC	: Herbier Des Comores
Ht	: Hauteur total
ISME	: Société Internationale pour les Ecosystèmes de Mangrove
MNHN	: Muséum National d'Histoire Naturelle de Paris
ONG	: Organisation Non Gouvernementale
PANA	: Programme d'Action Nationale d'Adaptation aux changements climatiques
PIB	: Produit Intérieur Brut
PNUD	: Programme des Nations Unies pour le Développement
PNUE	: Programme des Nations Unies pour l'Environnement
RFIC	: République Fédérale Islamique de Comores
RN	: Route Nationale
UICN	: Union Internationale pour la Conservation de la Nature
UNESCO	: Organisation des Nations Unies pour l'Education, la Science, et la Culture
WWF	: World Wild Fund

GLOSSARY

Anthropogenic: Refers to phenomena that are caused or maintained
by conscious or unconscious human action.

Backshore: The inner part of the tidal marsh between the mangrove and the mainland.

Biomass : Total mass of dry woody plant material per unit area.

Channel : **A** narrow natural or artificial passage between land and sea.

Buttress: Root expansions more or less vertically flattened, stuck to the trunk.

Boundary dune: Sand dune located behind lagoons or beaches.

Ecology : Science that studies the relationships of living beings with each other and
with the environment.
environment in which they live.

Endemic: Specific to a given geographical region.

Mataroma: Comorian term for "floor".

Halophilic environment: Salty environment.

Hypocotyl: Organ contained in the fruits of certain plants, ensuring germination on the mother plant
and thus giving the seedling.

Impact : **The** effect of a more or less severe disturbance on the environment or
on one or more elements.

Intertidal: Refers to that which is located between the highest and lowest tides.

Marsh : Low-lying area where stagnant water accumulates, characterized by
a particular vegetation and fauna.

Threat : All the parameters that could lead to the disruption, the
destruction of a given resource.

Nkassa : Comorian term for "dugout floaters".

Mangrove: Tree species native to tropical maritime marshes.

Landscape : **A** geographical area whose relief, climatic conditions, and
soils, vegetation and cultural elements are homogeneous.

Stand: A grouping of one or more species in a well-defined area.

Seedling : young plant resulting from the germination of the seed and which is
still developing from the cotyledonary reserves.

Pneumatophore: Ramification of the root system of certain mangroves. These roots are immersed
and vertical and ensure the aeration of the plant.

Pressures: A set of parameters that are permanently exerted on a given resource and lead to
negative changes in its natural state.

Phenology: Study of the influence of time and ecological conditions, including climate, on the
succession of the various phases of the life cycle of a species, especially a
higher plant.

Stilt roots: Roots developed by some mangroves from the trunk or branches. This adaptation allows
for better support, good ventilation and elevation above the water.

Sursalure : Salt accumulation in mangroves.

Tanne : Large area of bare or grassy soil located between the mangrove forest and
land.

INTRODUCTION

INTRODUCTION

For decades, an accelerated degradation of marine and coastal areas has been observed in the world, linked to coastal anthropization, climate change, and overexploitation of resources. The coastal environments of the Comoros Islands will probably not be spared if the rate of degradation accelerates.

The Comoros Islands are renowned for their rich marine and coastal ecosystems. Among these ecosystems, some are less studied and more threatened such as the mangrove ecosystem.

The mangrove is the whole of the vegetation which develops in the zone of rocking of tide of the intertropical littoral regions. It colonizes areas supplied with fresh water and sheltered from marine currents, such as estuaries and lagoon systems, i.e. calm and shallow areas. In tropical regions, it occupies nearly 75% of the coastline and deltas. According to estimates, it covers 14 to 23 million hectares worldwide (UNDP, 2010).

Today, mangroves are among the most threatened habitats in the world. The bitter fact is that the extent of these mangroves is decreasing more and more under the effect of demographic pressure, anthropic factors and natural factors. According to an FAO assessment (March 2003), the area covered by mangroves in the world has decreased from 19.8 million hectares in 1980 to less than 15 million in 2000. According to the same source, the total area of mangroves in the Comoros has been gradually decreasing since 1976. It went from 125 hectares in 1980 to 115 hectares in 2005 (FAO, 2005). This phenomenon is visible in some sites where the mangrove remains only as a trace.

However, an exact assessment of the situation of mangroves in Comoros is difficult to make due to the lack of previous studies in these environments. However, the degradation of this ecosystem is real. It should be noted that mangrove forests play an important role in the regulation of the greenhouse effect and the current situation of this ecosystem is becoming increasingly alarming. Thus, in view of the economic and ecological issues related to climate change, it is essential to study this environment and to know its role, especially its mitigation potential in global warming. It is in this context that the study of the present dissertation entitled: *"Ecological characterization of mangroves and evaluation test of carbon sequestration by mangroves: case of the island of Ngazidja" falls within* the global framework of the fight against climate change through carbon sequestration. It aims specifically to :

> Identify the ecological and biological characteristics of mangroves;
> To know the ethnobotanical and socio-economic uses of mangroves;
> Quantifying carbon sequestered by mangroves ;
> Propose alternative measures on mangrove management and climate change mitigation.

Thus, we have structured our study plan into four main parts:

- Study environment
- Methodology
- Results and interpretations
- Discussion and recommendations

PART 1:
STUDY ENVIRONMENT

STUDY ENVIRONMENT

I. GEOGRAPHICAL LOCATION

Located in the Indian Ocean at equal distance from East Africa and Madagascar (350 km), north of the Mozambique Channel, between 11°20' and 13°04' South latitude; 43°11' and 45°19' East longitude, the Comoros archipelago comprises four volcanic islands aligned over 225 km along a submarine plateau in a South-East/North-West direction, three of which: Grande- Comore or Ngazidja (1148 km2), Anjouan or Ndzouani (424 km2), Mohéli or Mwali (290km2), constitute the union of Comoros; the island of Mayotte or Maoré (370 km2) has remained under French occupation since 1975. The entire archipelago covers a total area of 2236 km2, the islands are about 30-40 km apart isolated from each other by deep underwater trenches along the African coast (Ramadhoini, 2011).

Ngazidja is the most volcanic, youngest, largest and most western island. It is located between 11°20' and 11°54' South latitude and 43°15' and 43°33' East longitude. It is 700 km from the northwest coast of Madagascar and 300 km from the east coast of Mozambique, 45 km from Moheli, 75 km from Anjouan and 180 km from Mayotte. In the central south, the island remains under the rule of Karthala, the volcano with the largest active caldera in the world (Andiliyat, 2007). (Map 1)

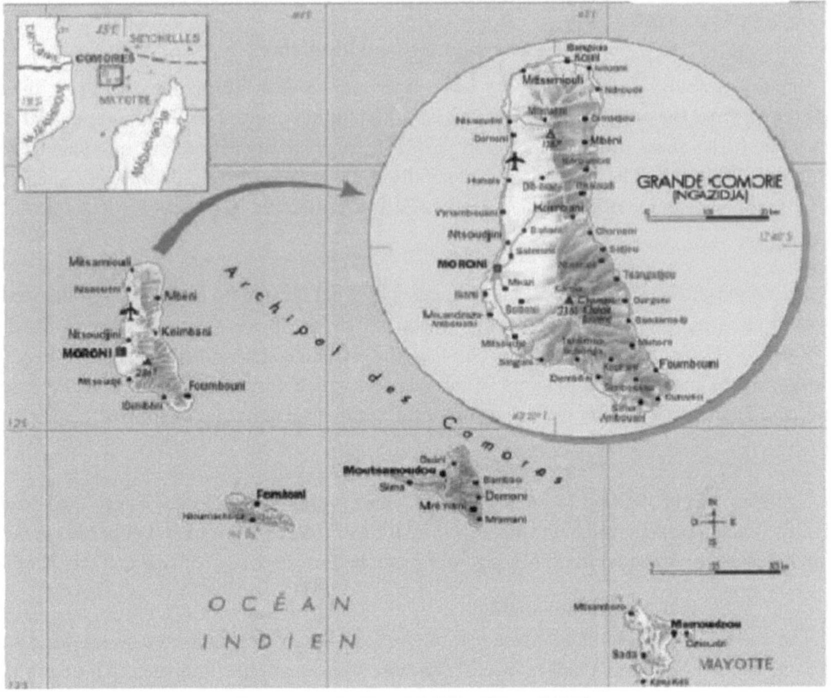

Source: WWW.quid (*//Andiliyat* 2007)

Map 1: Location of Ngazidja Island in the Indian Ocean and the Comoros Archipelago

II. ABIOTIC ENVIRONMENT

II.1. Climate

The climate of the Comoros is tropical and humid with an oceanic influence. The year can be divided into two main periods: a dry and cooler season from May to October and a humid and hot season from November to April. The insularity, the altitude, the irregularity of the reliefs are at the origin of a great diversity of local climates.

The island of Ngazidja presents two trends: a humid or very humid western region and a dry, sometimes very dry eastern coastal region. All intermediates between these two extremes are encountered (Andiiyat, 2007).

11.1.1- Temperature

The average annual temperatures are relatively constant throughout the year and vary on average between 25°C and 28°C at low altitudes. The maximums are observed in the rainy season and the minimums in the dry season. The temperature decreases by 0.7°C per 100m with altitude. On the coast, diurnal amplitudes are moderate during the rainy season, and the greatest are observed in June-July during the dry season (Ergo, 1984). They are certainly greater at higher altitudes, especially above 1000 m, but there are no statistics available on this subject (Battistini and Verin, 1984). (Appendix I)

11.1.2- Precipitation and insolation

Rainfall is abundant and highly variable from year to year. The winds that bring the rains generally come from the northwest. Rainfall varies with altitude: more than 5000 mm per year on the western slope of Karthala (the most watered part). The average annual rainfall in Ngazidja varies from 1500 to 5000 mm. The insolation is generally high, varying from 2000h/year to more than 3000h/year with an average of 2600h/year (Battistini and Verin, 1984). (Annex I).

11.1.3- Winds

The island of Ngazidja is subject to two types of wind depending on the time of the year: the south-east trade winds (Kussi), during the dry season and the north-west monsoon (Kashkazi) during the rainy season (Battistini and Verin 1969). In fact, the climate is determined throughout the year by the position of the major meteorological action centres, which correspond to areas of high or low pressure, and by the variations in the extension of the various air masses (Annex I).

11.1.4- Cyclones

Cyclones occur during the hot season. Three types of cyclones cross the Comoros episodically. Each type depends on the place of formation of the cyclone which is either in the vicinity of the archipelago, or in the North of Madagascar or in the East between 55° and 65° East longitude.

II.2. Geomorphology

The island of Ngazidja is elongated from North to South, and measures 64 km long and 24 km wide and has three mountain ranges, that of Karthala whose summit culminates at 2361 m altitude, that of the Grille, in the North of the island with an altitude of 1087 m and that of Mbadjini in the South whose summit rises to 650 m. The first two massifs are linked by the Dibwani pass which constitutes a passage at an altitude of about 500 m.

It has an area of 114,800 ha of which 40,000 ha is cultivable land, 10,800 ha is grazing land, 12,500 ha is forest massifs and 51,500 ha is non-cultivable land. Among its area of 1148 Km2, the forest area estimated at 12375 ha for the whole archipelago (UNEP, 2002.), had a reduction of 36% in Ngazidja for the year 2006, that is a loss of 5000 ha (Ramadhoini, 2011).

The geomorphology of Ngazidja (recent volcanism) is characterized by few white sand beaches: 4/6 of those that exist are found in the north of the island; the whole shows the presence of coral beds. About 90% of the island's sandy beaches have disappeared during the last decades, Map 2 (Andiliyat, 2007).

The mangroves of Ngazidja are not very developed, and are located especially on the west coast of the island, notably in Ikoni, Moindzazamboini, Voidjou, Hahaya, Domoimboini, in the South they are found in Ouroveni, Chindini, Simamboini, Mohoro, and in the North they are located in Seleani, Ouellah, Chomoni. (UNEP, 2002 and personal data).

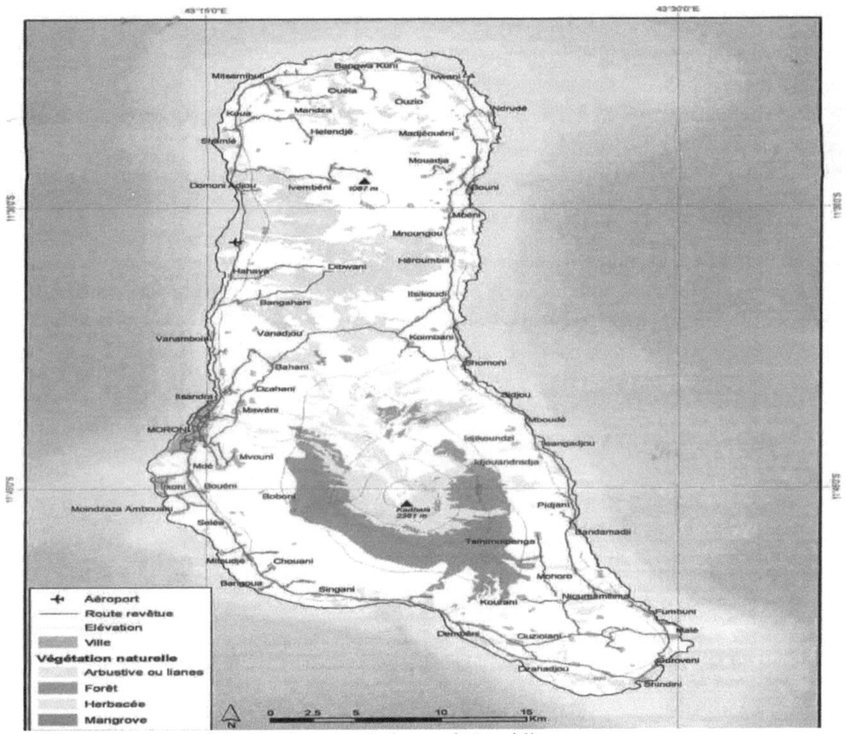

Map 2: Geomorphology of Ngazidja

II. 3. Hydrography, and substrate

The hydrographic network of Ngazidja is non-existent because of the volcanic porosity. The mangrove is essentially found on loose substrates, muddy to varying degrees.

The soil of mangroves is generally characterized by the presence of a permanent water table, with a high sodium content and sulphate-reducing conditions linked to the anaerobic environment (Andriamalala, 2007). Some mangroves in Ngazidja are installed on substrates characterized by volcanic rocks.

III. BIOTIC ENVIRONMENT

III. 1. Vegetation and flora

111.1.1- Vegetation

Several types of vegetation have been distinguished in the Comoros according to altitude: dense humid forests, shrubby and bushy thickets, savannahs (shrubby and grassy), mangroves, swamps, marshes and grasslands, saxicolous groups on slag, plantations and crops (Adjanohoun, 1982).

The mangrove is part of the dense forests of low altitude. It is a plant formation with a very complex ecology. In general, it consists of an evergreen forest established on the flat, muddy or rocky shore of the sea. If we speak only of the surface occupied, the substrate and the vegetation, then we distinguish from the sea to the land:

❖ The coastal dunes parallel to the coast;
❖ The mangrove itself is characterized by the species of mangroves;
❖ The back mangrove which is a vegetation formation characterized by trees, shrubs, and herbaceous plants.
❖ The tanne which is a large area generally bare. However one can meet a grassy tanne.

111.1.2- Flora

The flora of the Comoros Islands is estimated at more than 2000 species (Adjanohoun et al., 1982). According to the same author, nearly 500 species have been identified in the Comoros and are distributed in 106 families and 350 genera. For the whole vascular flora, more than 33% of indigenous plants are endemic.

The flora of the mangroves in Ngazidja is poor in species because of the severe conditions in which some plants have managed to survive. It shelters 7 species of mangroves distributed in 6 families. (Tableaul and Plate 1).

Table 1: Mangrove species in Ngazidja

Family	Genus and species	Vernacular name
AVICENNIACEAE	*Avicennia marina*	Mhonko mwewu
COMBRETACEAE	*Lumnitzera racemosa*	Mhonko mché
MELIACEAE	*Xylocarpus moluccensis*	Mfubu
MALVACEAE	*Heritiera littoralis*	Mhonko
RHIZOPHORACEAE	*Bruguiera gymnorrhiza*	Mhonko chouma
	Rhizophora mucronata	Mhonko mmé
SONNERATIACEAE	*Sonneratia alba*	Mhonko mouigni

Source: Personal data

III.2. Wildlife

The number of species of fauna of the Comoros Islands varies according to the authors. Many species have been recorded, some of which are endemic:

- ❖ 17 species of Mammals with 2 species and 3 varieties are endemic, including *Pteropus livingstoinii* and *Eulemur mongoz* (the Comoros maki), (Mbae et al, 1991)
- ❖ 101 species of birds including 35 endemic varieties (14 species forming an endemic genus). (Louette et al, 2004)
- ❖ 1106 species of insects (Decelle, 1980)
- ❖ 25 species of Reptiles (including turtles) of which 11 are endemic
- ❖ 32 species of freshwater fish and macro-crustaceans, of which 2 are endemic, and nearly 820 species of marine fish, including *Latimeria chalumnae* (Keith et al 2006).

In the case of the Comoros mangrove, it is characterized by an exceptional and abundant fauna but poor in species. The groups of animals often encountered are: fish, crustaceans, molluscs, reptiles and birds.

Plate 1: Existing mangrove species in Ngazidja

III.3. Man and his activities

III.3.1- Settlement History

The first traces of settlement date back to the 8th century and the Comorian population is of Bantu, Arab/Chirazian, Indonesian, Malagasy origin, etc... (Map 3).

The Comoros Islands have been populated by successive waves of migrations from the

Persian Gulf and East Africa and have been enriched more recently by exchanges with the Malagasy population. In spite of its diverse origins, the population is characterized by a great homogeneity and religious (Muslim), linguistic and cultural unity.

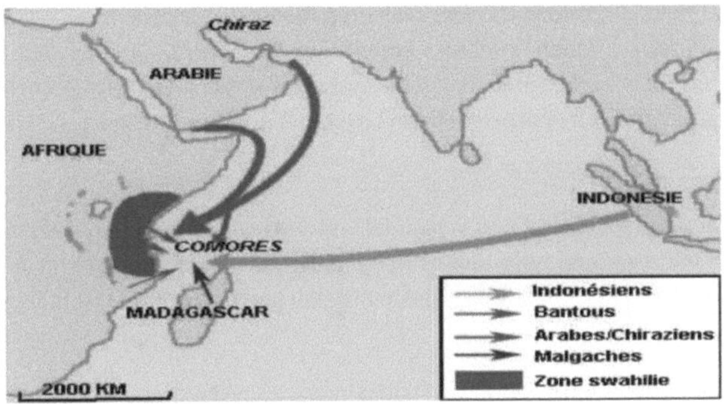

Map 3: Origin of the Comorian population (source malango-Comores, 2006)

III.3.2-Demographics

According to the latest census data from the Direction Nationale du Plan (2007), the Comorian population continues to grow at a rate of 2.7%. The latest figures for the population of the Comoros show a total of 892182 inhabitants in 2006, i.e. an average annual density of 335 inhabitants per km2. This population growth is evolving in parallel with the pressure on natural resources. (Table 2)

Table 2: Estimated demographic profile of the Comoros in 2002

Indicators :	Workforce
Growth rate	2,7%
Sex ratio (men per 100 women)	0,49
Density	335
Population under 25 years old	56%
Urban population	30%
Rural population	70%
Population in the coastal zone	65%
Population residing outside	35%
Life expectancy at birth	56.5 years old

Source: National Planning Directorate

In Ngazidja, the population is 293,545 with a growth rate of 2.0% and an annual density of 286.6 inhabitants/km2 (Map 4).

III.3.3- Population activities

The main economic activities of the Comoros are based on agriculture, livestock and fishing. They constitute the primary sector and contribute to about 40% of GDP and all exports. Exports of vanilla, ylang-ylang and cloves account for 90% of the country's exports.

> *Agriculture*

Agriculture occupies more than 70% of the active population and corresponds to 60% of the GDP. The area reserved for crops associated with natural vegetation is 42,200, of which 32,700 ha are for cultivation and 9,500 ha for natural vegetation. The main activity of the population is agrosylvo-pastoralism, which provides for the needs of wood and agricultural products.

There are various agricultural systems in Comoros, including: open field food crops, food crops under natural forest, agroforestry, and cash crop monoculture. The most common system used in Ngazidja is agroforestry, which combines herbaceous food crops, shrubby cash crops, fruit trees and sometimes forest trees in the same plot. This system is stable and varies according to ecological conditions. It is characterized by permanent soil cover and good use of space. (DGE, 2002).

> *Breeding*

Livestock activities are practiced in parallel with agriculture but in a poorly organized manner. They contribute to the meat and milk diet of the population. These activities mainly concern cattle (about 43,000 head), goats (113,000 head), sheep (18,000 head), poultry (160,000 head), and donkey breeding in Moheli (Abdillah, 2009).

> *Fishing*

Comoros has an under-exploited potential in terms of fisheries resources. The number of fishermen is estimated at 7000. Among the techniques used are :
- The traditional and artisanal fishing which is practised by means of outrigger canoes made of forest wood. The number of pirogues is estimated at 4400 units for the whole archipelago;
- Fiberglass boats manufactured locally. These motorized boats, estimated at 487 units, promote fishing efficiency;
- Foot fishing at low tide is practiced mainly by women and children and flatfish fishing is done with a harpoon;
- The dynamite fishery, which, despite its ban, still takes place (DGE, 2002).

III.4. Socio-economic importance of the mangrove

The socio-economic importance of the mangrove in Ngazidja can be deduced from the products it can provide. Indeed, mangroves provide various resources such as firewood, construction wood, timber, medicinal plants, fishery products (fish, shrimps, crabs). It is also a place of grazing for livestock, a place of leisure, eco-tourism and a place of sharing traditional customs.

The absence of a management system and legislation on mangrove preservation in Comoros means that the devastating action of man is beginning to be felt in recent years.

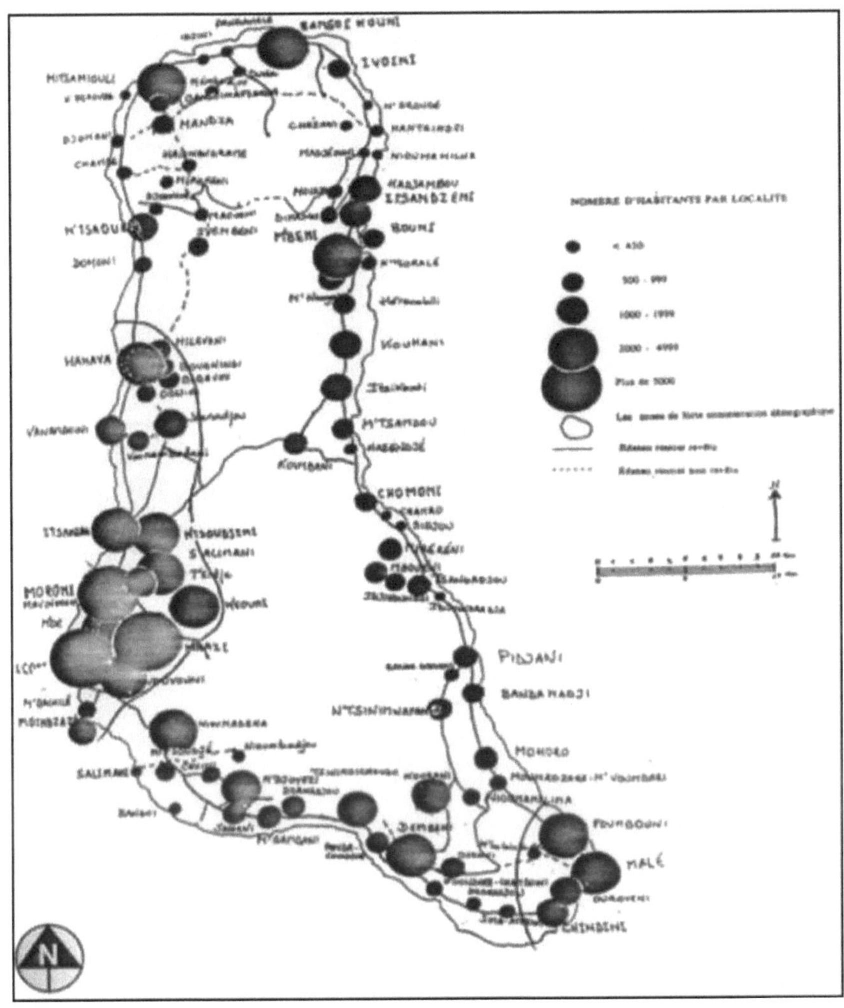

Map 4: Population of Ngazidja (Source: Andiliyat, 2007)

PART 2:
METHODOLOGY

METHODOLOGY

The objective of this work is to make an ecological and socio-economic study of the mangrove ecosystem of Ngazidja and an evaluation of the carbon stock. To do this, we adopted several methods:

- ❖ Bibliographic research ;
- ❖ Ethnobotanical surveys ;
- ❖ Ecological surveys ;
- ❖ Calculation of biomass and carbon stock.

I. BIBLIOGRAPHIC SEARCHES

They consist in collecting data before and after the field trip. In fact, all the documents related to the theme and the study environments were consulted before the field trip, namely

- ❖ information on the environment and biodiversity of the Comoros ;
- ❖ general and specialized works concerning the works already carried out on the mangroves of Comoros, Madagascar and other countries;
- ❖ official websites.

All the necessary information and archived research related to the topic was collected. This is especially true of conceptual definitions, methodologies and techniques as well as publications of the results of previous work. These bibliographical supports helped us to write this report of study dissertation as well as possible. The references of the consulted documents are reported at the end of this manuscript.

The documentation centres frequented are those of :

- University of Comoros (library, herbarium) ;
- DGEF (Directorate General for the Environment and Forestry) ;
- CNDRS (National Centre for Documentation and Scientific Research).

II. ETHNOBOTANICAL SURVEYS

The ethnobotanical survey consists of the study of the relationships between plant species and riparian populations. The objective of the survey is to have information on the typology of use of mangroves and their economic values.

11.1. Selection of survey sites

These sites were chosen based on the proximity of the villages to the study areas. Seven villages were the targets of our survey: Ikoni, Batsa, Domoimboini, Mohoro, Ouroveni, Seleani, Ouellah.

11.2. Survey modes

Two types of approach were adopted: collective (group) and individual surveys. They were carried out in the form of closed, semi-open or open questions. The questions and answers were recorded on survey sheets as the interview progressed. The questionnaire focused on the vernacular name, the mode of use, the importance of each species, the part used, the mode of preparation... (Annex II). (Appendix II).

The different people who responded to the interview were: farmers, fishermen, masons, notables, housekeepers, students. The age criteria were not neglected: young, adult, old.

II. 3. Species Utilization Index

In order to know the most used species in mangroves, we proceeded to the calculation of the use index for each species. It is determined from the formula of LANCE et al (1994). The formula is as follows:

$$I(\%) = n / N \times 100$$

With **n**: number of people citing the species.

N: number of people surveyed.

In our case we have considered :

- ❖ **n** : number of villages where the species exists.
- ❖ **N**: number of villages surveyed.

Since most of the people interviewed do not know the names of the species but consider that all mangroves existing in the mangrove are used.

The information obtained during the surveys allowed us to evaluate the use of the selected species. Indeed, according to the value of this index we can conclude that :

- A plant is well known and widely used if its index varies between 60 and 100%.
- A plant is moderately known and used if its index varies between 30 and 60%.
- A plant is little used and probably little known if its index is lower than 30%. (LANCE et al, 1994).

III. ECOLOGICAL SURVEY

III. 1. Direct observation: Presence - Absence (fauna and flora)

Direct observations have been used for all types of diagnosis. Together with photography, they clarify all situations:

- ❖ Populations and vegetation types.
- ❖ Threats and pressures.
- ❖ Biodiversity potential (fauna and flora).
- ❖ Changes to the ecosystem and habitats (state of degradation).

This method gives a prediction of the adequate solutions for the observed situations.

For fauna, the direct observation method with the naked eye was used. We proceeded to distant observations for birds and close observations for other animals (crabs, fish, shrimps). Images were taken to facilitate species identification. The identification of crab and mollusc species encountered in the mangrove was done using the works of Richmond, 2002 and Louette et al 2004.

III.2. Method for studying vegetation

111.2.1- Preliminary survey and site selection

The preliminary survey aims to obtain a general knowledge of the study area and its environment. It is a phase for formulating hypotheses and identifying inventory sites. The survey also consists of identifying the type of land use and occupation in order to better assess the pressures on the study area.

The choice of study sites was made on the basis of bibliographical information on the ecology, floristic richness, geographical location, state of degradation, pressures and threats, as well as the socio-economic value of mangroves. On the other hand, the choice of the location of the field surveys depended on the methodology applied and the study environment.

111.2.2- Duvigneaud Transect Method

For each floristic survey, we carried out a transect that crosses the different plant formations.

The objective of this method is to characterize the different types of plant formations and to study the floristic zonation of each mangrove. These transects of variable length, are parallel to the greatest slope and perpendicular to the sea (figure 1).

The search for the minimum area consists of recording the species present in each elementary area and noting the one that appears each time the area is doubled. The minimum area is the smallest area where the maximum number of species is found.

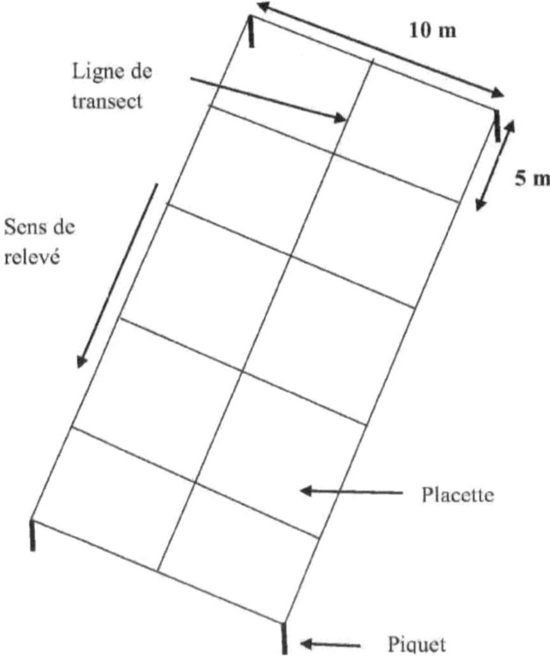

Figure 1: Transect layout according to Duvigneaud's method
III.2.3- Placeau method

Given the homogeneity and floristic zonation in mangroves, we also adopted the plot method in order to identify the floristic richness and density of the plant formation. A plot is a sampled area of any shape but we generalized it to a square shape.

This method is used on a floristically homogeneous surface if the surface does not show any appreciable difference in floristic composition between its different parts. (Gounot, 1969). It consists of making an inventory of the floristic species present while taking into account the following homogeneity criteria:

- ❖ Uniformity of apparent ecological conditions ;
- ❖ Physiognomic homogeneity ;
- ❖ Homogeneity of the floristic composition.

Survey area

For each plant formation type, the survey area is a 10 m square. Each square or plot is then subdivided into four segments or plots of 5 m sides. The selected plot is marked with a wire 1 m

21

above the ground, fixed at its four ends. Each segment is then marked with stakes.

Reading device

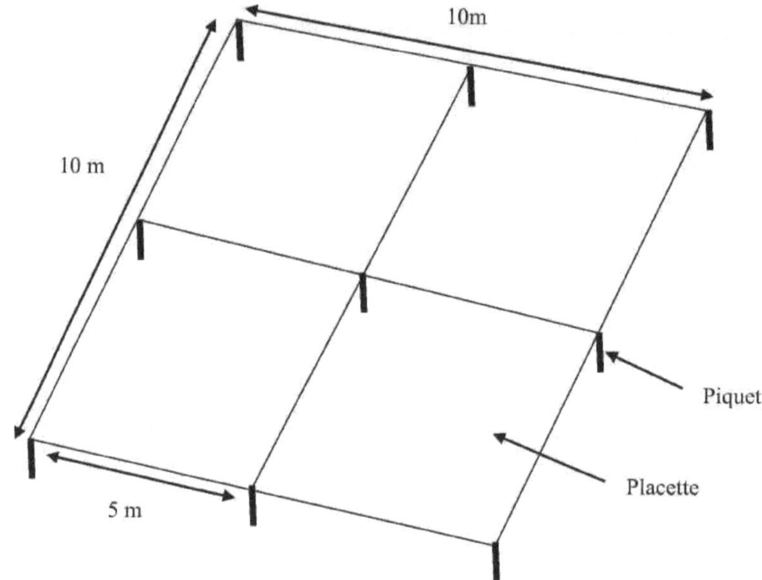

Figure 2: Placeau model according to the Braun Blanquet method
111.2.4- Some field materials

A digital camera to preserve some of the details that can be used for illustration and to facilitate the description of the features;

A GARMIN MAP 76 GPS, a satellite navigation tool, for geo-location of survey sites;

A compass with an ergonomic plate for transect orientation;

A 50 m decameter and **a rope** to delimit and set up the transects and plots;

1.5 m long graduated **tape measure** for measuring DHP ;

A pruning shears used for collecting the specimens. The fertile samples (presence of fruits, flowers, grains) were collected in priority and the sterile samples are collected in the case of absence of fertile individuals;

A plastic bag required for the conservation of the samples;

A weed press with newspapers to spread out the samples from each other to keep them in good condition during drying;

Ecological survey sheets in which the ecological and floristic parameters are shown (Appendix III).

111.2.5- Vegetation structure and zonation of the mangrove

> The structure of the vegetation is the way in which the plants are arranged and distributed in relation to each other. (GUINOCHET 1973). In our case, we have considered two levels, each corresponding to a stratum. Each stratum is a function of height:

- An upper stratum occupied by adult individuals;
- A lower stratum occupied by young plants.

> The zonation of the mangrove: It is the distribution of plant species on the surface of the ground. That is to say that the distribution is done by a zonation of the species going generally from the sea to the land. This method is based on the establishment of a schematic profile of a part of the vegetation as representing the whole formation studied.

111.2.6- Vegetation study parameters

Survey sheets were prepared for each field trip. In each sheet, we mentioned the ecological parameters, the floristic parameters and the state of health of each mangrove.

a. *Ecological parameters*

❖ Statement date ;
❖ Location;
❖ Geological coordinates (altitude, latitude, longitude) ;
❖ Orientation;
❖ Survey size ;
❖ Stratum of belonging.

b. *Floristic parameters*

In the survey column, the scientific names of each species encountered along the transect or in the plot are given. On the line are placed the following parameters:

> Diameter at breast height (DBH) of an individual for a species in a given stratum: this is the diameter of a tree or shrub at breast height or at 1.3 m from the ground or above the stilt roots and buttresses;

> Total height (Ht) of an individual for each species in a given stratum;
> Numerical abundance: total number of individuals per species present;
> The phenological state of the species (fruiting, flowering, leafing, vegetative).

c. *Health status*

In our survey sheets, we have mentioned the health status of each individual. Depending on the case, the state of health is Vigorous (if the tree shows no trace of cutting), Normal (if the tree shows traces of cutting), and Damage (if the tree dies after cutting).

a) Délimitation du transect à l'aide d'une corde

b) Division du transect en placette

c) Estimation de la hauteur d'un arbre

d) Mesure de DHP à l'aide d'un mètre ruban

Plate 2: Setting up the survey devices and measuring the floristic parameters

111.2.7- Study of the natural regeneration of mangroves

This study is based on the counting of seedlings and adult plants in each survey within each mangrove type.

a. *Natural regeneration method*

Mangroves regenerate by the dissemination of seeds and seedlings. According to the germination and dispersion of seeds and seedlings, mangroves are grouped into two categories: viviparous and non-viviparous species.

> **The viviparous species** have seeds that germinate on the mother plant. At maturity, the seedling detaches itself from its parent and settles in the mud where it takes root and continues to grow; these are the RHIZOPHORACEAE. The size of the seedlings varies according to the species and the environmental conditions.

> **Non viviparous species**, such as *Avicennia marina, Sonneratia alba, Xylocarpus moluccensis, Heritiera littoralis,* whose organs of dissemination are carried by water. The fruit of *Avicennia marina* is cordate and slightly compressed with a persistent style and calyx that allows it to float easily on water. The fruit of *Xylocarpus* is spherical and floaty, requiring a strong obstacle to germinate. The obstacle will break the pericardium and the seeds will be released. That of *Heritiera littoralis* is provided with a keel or a small veil being used as float. That of *Sonneratia alba* floats

thanks to the calyx. When this fruit sinks to the bottom, it germinates.

b. Rate of natural regeneration

The results on natural regeneration are obtained only from the species encountered in the survey plots. To calculate the regeneration rate, we count the regenerated individuals (DBH <10cm) and the seed individuals (DBH >10cm). The parameter considered is the diameter at breast height (DBH).

The regeneration rate (**RR**) **is** the ratio of the number of seedlings to the number of mature individuals per station multiplied by 100. (ROTHE, 1964):

$$TR\ (\%) = Nr\ /\ Ns\ x\ 100$$

Nr: Number of regenerated individuals (DBH < 10 cm)

Ns: Number of seed individuals (dbh > 10 cm)

RR: Regeneration Rate

Depending on the value of the regeneration rate, three cases are possible:

- ❖ if TR < 100 %: the regeneration is weak;
- ❖ if 100% < TR < 300%: regeneration is good;
- ❖ if TR > 300%: the regeneration is very good.

111.2.8- Collection and identification of species

All mangrove species and associated plants present in each ecological survey were collected and placed in a herbarium. Species determination was carried out at the laboratory of plant biology and ecology (Comoros herbarium) of the University of the Comoros by a systematist or with the help of a software called "determination key for woody plants of the Comoros" (Photo: Appendix V).

IV. CALCULATION OF BIOMASS AND CARBON STOCK

IV.   1. Calculation of the biomass

In order to evaluate the rate of carbon sequestration for each mangrove, we calculated the biomass of each one. From this biomass, the carbon stock of each site can be determined.

The calculation of biomass is based on the measurement, by biological inventory, of the dbh and height of mangrove trees. Biomass corresponds to the mass of dry woody plant matter per unit area. It is determined from the FAO formula (1990). The parameters considered are the diameter at breast height (DBH > 10 cm) and the maximum height for each site. The formula is as follows:

$$BA = 0{,}544.\ [\textstyle\sum(DHP)]^2.\ Ht$$

BA : Biomass rate in Kg

DBH: Diameter at Breast Height > 10 cm

Ht: Maximum height of a mangrove tree per site

In each site we considered a plot of 100 ^{m2} equivalent to 0.01 ha. So after calculating the biomass, we multiplied its value by 0.01 to get the value in Kg/ha.

IV.   2. Calculation of carbon stock

The calculation of the carbon stock is obtained by multiplying the biomass value obtained by a conversion factor of **0.5** (FAO, 1990).

$$C = 0{,}5.BA$$

C : carbon stock in Kg / ha

<u>*NB*</u>: Herbaceous and shrubby strata, non countable trees and dead wood have not been taken into account in our calculations.

The information obtained during the ecological surveys of mangroves allowed us to estimate the rate of carbon sequestration in each site but also to evaluate the importance and the exact role of mangroves in the regulation of the greenhouse effect.

V.   MAPPING STUDY

Mapping will be carried out after the field work, through :

- ❖ Topographic map of Ngazidja ;
- ❖ Data recorded in the field in 2012, using a GPS ;
- ❖ Data analysis using MapInfo 7.5 software.

PART 3:
RESULTS AND
INTERPRETATIONS

RESULTS AND INTERPRETATIONS

I. RESULTS OF THE ETHNOBOTANICAL SURVEYS

The respondents are generally male. Most of the women refused to answer our questions. The age of the respondents varies between 18 and 80 years, of which 90% are elderly.

The results of the surveys allowed us to know the use of the species and the pressures they undergo. Several species from the mangrove are used in a traditional way by the villagers.

Surveys were carried out in seven villages near the study sites. A survey form concerning the species of wood used was established (see Annex II). A total of 104 people were surveyed, including 14 in Ikoni, 13 in Voidjou, 10 in Domoimboini, 23 in Mohoro, 15 in Seleani, 18 in Ouroveni, 10 in Ouellah.

I.1. Use of species

The use of mangroves and associated species is common among the traditional population. Indeed, these species are used in various forms depending on the locality and the needs. In various sites, the mode of use is the same. During the ethnobotanical surveys, several species used by the local population were identified. Table 3 shows the traditional use of some mangrove species.

Table 3: Use of woody resources in mangroves

Families	Scientific names	Vernacular names	use
	Rhizophora mucronata	Mhonko mmé	Bch, Bc, Co, So, F,
RHIZOPHORACEAE	*Bruguiera gymnorrhiza*	Mhonko shuma	So, F, Bch, Co, med, Cs
SONNERATIACEAE	*Sonneratia alba*	Mhonko mouigni	So, Bch, Bc, Jx, med, Bo
AVICENNIACEAE	*Avicennia marina*	Mhonko mewu	So, Bch, Bo, med
COMBRETACEAE	*Lumnitzera racemoza*	Mhonko mché	Bch, Bc, So, med, Cs
MELIACEAE	*Xylocarpus moluccensis*	Mfubu	Bc, Bo, Bch,
MALVACEAE	*Heritiera littoralis*	Mhonko	med, Bch, Bc
RUBIACEAE	*Guettarda speciosa*	Foulamboi	med, Bch, Cs, Ap
LYTHRACEAE	*Pemphis acidula*	Mdjana	Bch, So
MALVACEAE	*Hibiscus tiliceus*	Mwaro	Bch, Bo,
FABACEAE	*Caesalpinia bonduc*	Mtso	Jx, med
EBENACEAE	*Euclea* sp	Mlala	Bch, Cs
FABACEAE	*Tamarindus indica*	Mhadju	Al, Bc, Bch, Bo

Sources: Personal data

Co: dye M ed: medicinal plants Bch: firewood
Bo: lumber Al: food Cs: cosmetics
So: witchcraft Bc: timber F: fodder
Jx : games Ap : aphrodisiac

1.1.1- Firewood

Mangroves are among the species used for fuelwood, especially for villagers living in coastal areas, given their proximity to mangroves. In Ngazidja, fuelwood is used extensively in rural areas especially during the month of Ramadan, when demand for fuelwood increases. Dead wood is most commonly used, with some villagers preferring mangroves because of their energetic power which reduces cooking and collection time. The frequency of fuelwood collection is once a week. Rhizophoraceae are the most popular. However, the other species represent very poor quality firewood.

1.1.2- Construction timber

Mangrove trees are used by the local communities for the construction of houses, huts and fences. They are mainly used to build "valla" huts for boys forced to leave their parents' home.

Rhizophora mucronata is the most used species because of its strong and straight trunk. This species is used as a roof support and as a pillar.

Lumnitzera racemosa is also used as a pillar in house construction. *Xylocarpus moluccensis* and *Bruguiera gymnorrhiza* can also be used in construction but they are less appreciated because of their non-straight trunk.

The other species are used to make fences for fields and sheep pens.

1.1.3- Lumber

The woods used for the manufacture of dugouts are large diameter woods, easy to work, light and resistant to sea water. The most sought after species are *Avicennia marina, Sonneratia alba, Xylocarpus moluccensis*. These species are used as crossbeam or support inside the dugout canoe called "mataroma" or used for the manufacture of board and floats "Nkassi" of the dugouts.

Some mangroves are used to make wooden beds and tables. The pneumatophores of *Sonneratia* are used to make the ends of table pillars.

1.1.4- Medicinal plants

Mangroves have several therapeutic virtues. Their use in healing many diseases is a widespread practice in traditional medicine. The parts used from mangroves are usually the fruits, leaves, roots or bark. Riparian communities use both mangroves and plants associated with mangroves (Table 4).

Table 4: Effects and therapeutic virtues of mangroves and associated species

Species	Parts used	Diseases	Preparation	Dosage
Avicennia marina	Sheets	Urine control problem in children	Boil in water	To be drunk regularly
	Stems	Against mosquitoes	To burn	
	Resin	Toothache	Casting the resin from the trunk	To put on diseased teeth
Bruguiera gymnorrhiza	Sheets	Knee pain	Boil the leaves	Massage the knees with the warm leaves
Caesalpinia bonduc	Sheets	Abortifacient, abdominal pain	Boil the leaves	To be drunk regularly
Dodonea viscosa	Sheets	Calcium deficiency and contraction	Boil the leaves	To be drunk regularly
	Sheets	Against mosquitoes and lice	Leave the leaves under the bed	
Guettarda speciosa	Sheet	Stomach aches	Boil in water	According to the needs

	Flowers	Postpartum infections	Boiling the flowers	To be drunk twice a day
Heritiera littoralis	Fruits	Boils Wound	Crush and apply the almond to the pimple or wound	Morning and evening until symptoms disappear
Ipomea pes caprae	Sheets	Joint problems	heating the leaves to a low temperature	Massage the joints with the warm leaves
	Sheets	Painful periods, Vaginal infections	Boil the leaves	Drink the infusion as needed
Lumnitzera racemosa	Barks	Stomach aches	Boil the bark	Drink one teaspoonful once a day
	Sheets	Malaria	Boil the leaves	Heating with steam (fumigation)
Morinda citrifolia	Fruits	Stomach ache, bloating	Blend the fruit	Drink the juice twice a day
Sonneratia alba	Sheets	Stomach ache	Boil in water	To be drunk regularly

Sources: Personal surveys and documentation

The colouring of traditional clothes and shoes with the bark of mangrove trees is a fairly common practice in communities living near mangroves. At the time, having a boubou[1] and shoes colored with the ink of Rhizophoraceae was a prestige for the great notables.

The bark of *Rhizophora mucronata* and *Bruguiera gymnorrhiza* is used to colour white boubous (biliguidji) and traditional cowhide shoes (zilatru). In fact, the bark is crushed and then soaked in water with the materials to be coloured for 2 to 5 days.

Mangroves and accompanying species are used as cosmetics for facials.

The flowers of *Lumnitsera racemoza* or *Guettarda speciosa* and *scaevola sericea* are crushed and used as make-up for women. The roots of *Guettarda speciosa* and *Bruguiera gymnorrhiza* are scraped and applied as "Msindzanu" mask on the face. The leaves *of Ipomea pes caprae* are crushed and used as shampoo. The roots of *Euclea* sp are used as a toothbrush to whiten teeth and stain lips. The fruit of *Xylocarpus* is used as a face mask by some women.

Mangrove trees are used in witchcraft as medicinal products[2]. The mangrove is a place of worship and remedy collection. Indeed, the use of mangroves in witchcraft is very common among some witch doctors. If a person is haunted by evil spirits "djinns" the sorcerer boils a mixture of mangrove leaves, then makes the patient inhale the steam: the smell and the gas make the evil spirits flee. The roots of *Sonneratia alba* are used to perpetuate couples' relationships when they are in distress. They are also used to cure male impotence problems.

---

[1] Traditional dress for men
[2] Pre-inscribed remedy by wizards

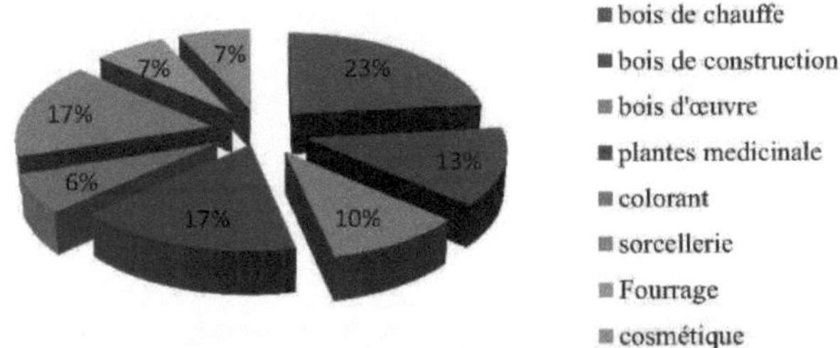

Figure 3: Use of mangroves

a) Entassement de bois de chauffe dans le sol

b) Clôture de champ par des palétuviers

Plate 3: Use of mangroves

I.2. Utilization index of species used

The surveys carried out show that all the species present on the mangroves are very much used by the riparian community. Their use index varies between 14, 42% to 100%. The species with an index of use exceeding 30% are known, and the most used have an index between 60% and 100%.

Table 5: Utilization index by species

Species	Use	Usage index
Avicennia marina	Medicinal plants, timber	22,11%
Bruguiera gymnorrhiza	Construction wood, dye, firewood, medicinal plant	68,26%
Rhizophora mucronata	Timber, dye, firewood	41,38%
Sonneratia alba	Lumber, firewood, medicinal plants	50,96%
Xylocarpus moluccensis	Construction wood, firewood,	24,03%
Heritiera littoralis	Medicinal plant, firewood,	14,42%
Lumnitzera racemosa	Construction wood, firewood, medicinal plant	77,88%
Guettarda speciosa	Medicinal plant, firewood, cosmetic	46,15%

(Source: personal surveys)

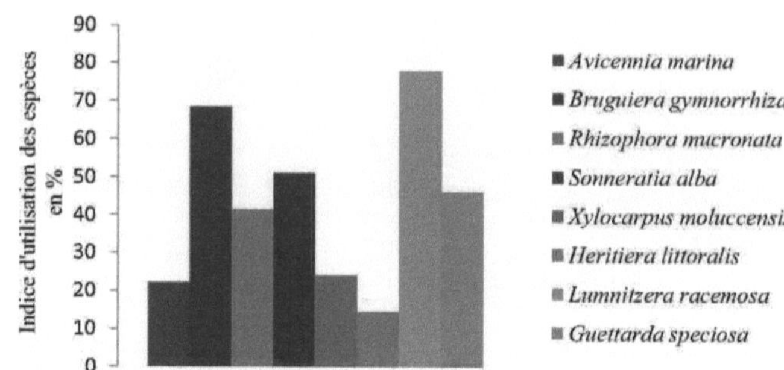

Figure 4: species use index

Lumnitzera racemosa and *Bruguiera gymnorrhiza* are the most used species with respective indices of 77.88% and 68.26%. Since they have a wide distribution and are very abundant and used in several areas. *Rhizophora mucronata* and *Sonneratia alba* have lower indices because they have a restricted distribution but are very abundant and are used in many areas of daily life by farmers. *Heritiera littoralis* has a very low utilization index (14.42%), as this species is in small quantities and is not widely used.

Partial conclusion

Our studies show that the mangrove of Ngazidja is under the influence of many factors responsible for its degradation, more particularly wood cutting. At the end of our investigations, we found that the use of wood in the construction of houses cannot harm this ecosystem because the quantity taken is still low. This harvest remains low, as most Comorians prefer to build their houses in brick or sheet metal (with forest wood). On the other hand, the consumption of firewood is quite alarming because most households in the villages surrounding the mangroves generally use mangrove wood.

II.   BIOLOGICAL CHARACTERISTICS OF MANGROVES

The ecological conditions in the mangrove ecosystem are very severe. Mangroves have settled on a halophilic environment. Its high salt content becomes a selective character, which confers a floristic poverty. The different species of mangroves adapt to the soft and asphyxiating substratum by emitting a rooting system through stilt roots, pneumatophores or buttresses according to the species. Evergreenness is total for all mangrove species (Michel, 1995).

II.1. Description and ecology of mangroves

a. Bruguiera gymnorrhiza: This dominant species is found in most of the sites studied, with the exception of Seleani and Ouroveni. It is found in dense and sometimes mono-specific stands. Its height varies between 2 and 10 meters. It presents a straight trunk without any major branching with kneaded roots, the fruit has an ovoid and short shape compared to *Rhizophora mucronata*. It has an affinity with *Avicennia marina* and *Sonneratia alba*. It does not form pure and extensive stands and prefers sandy-muddy environments.

b. Rhizophora mucronata: It is a species that is generally straight, its height varies between 2 to 12 meters. It differs from *Bruguiera* by its stilt roots in the shape of a semicircle that can measure up to 3 meters long and its leaves with a mucron at the end. It is a very demanding species from the point of view of pedology and salinity of the substrate. It stabilizes in soft mud and prefers clayey-silty soils with a foul smell. It often forms a dense, mono-specific stand that is difficult to penetrate in the internal zone of the mangrove.

c. Avicennia marina: It is a pioneer species with radiating roots equipped with a multitude of pneumatophores. The latter increase the plant's surface area and contribute essentially to the plant's respiration through their lenticels.
Its height varies from 3 to 7 metres. It is located from the sea to the land and it is less abundant compared to the other species because it is very severe and more demanding on the ecological conditions (daily immersions, muddy substratum, salinity). However, in Ngazidja it is found at the edge of the sea or inside the mangrove but it forms heterogeneous and discontinuous populations.

d. Sonneratia alba: is typically a sedimentation front species. It has a radiating root system with large, rigid, cone-shaped pneumatophores. The leaves are rounded and crassulescent. Its height varies between 3 and 7 meters. It is a pioneer species that colonizes silty or sandy alluvial deposits. It settles especially in the intertidal zones and can constitute a mono-specific formation along the littoral or behind micro-dunes. In Ngazidja, it is found in the fronts of lagoons and sea fronts of coastal environments. It often forms a mixed stand with Rhizophoraceae.

e. Lumnitzera racemosa: This widely distributed species is identified by its small leaves and the almost permanent presence of flowers, with a height varying between 1 and 7 meters. It is often in association with other species whose vicinity it supports. It is rarely found in the mangrove itself, and is located either behind the mangroves or behind the tans. It prefers areas with lower humidity and low salinity. It does not form a continuous formation but often a group of a few individuals and is found in all the sites studied except Mohoro.

f. Xylocarpus moluccensis: It has a very restricted distribution. It is a branched tree with laterally compressed winged buttresses. Its length varies between 2 and 8 meters high. It is characterized by its large spherical fruits. It settles in lightly salted and less flooded environments. It does not form a pure stand but mixes with other mangroves. It is often found on dry land.

g. Heritiera littoralis: This species with a restricted distribution is less abundant. Its length can reach 10 meters high. It can be recognized by its elliptical leaves, silvery on top and by its fruits characterized by a median ridge. It has a very high wood potential. This species is found behind the mangrove on less salty soil.

_NB_: The species *Ceriops tagal* of the Rhizophoraceae family is part of the mangrove species in Comoros but it was not encountered in our study sites.

II.2- Phenology of mangroves

The season determines the phenology of the species. Our field trip (June to September) coincided with the fruiting and flowering period of *Sonneratia alba* while *Lumnitzera racemosa* only flowers during this period.

The period of observation corresponds to fruiting phases of *Xylocarpus moluccensis, Bruguiera gymnorrhiza, Heritiera littoralis* and *Rhizophora mucronata,* while *Avicennia marina* was in vegetative state during our entire field trip.

In Ikoni, we observed defoliation in some individuals of *Xylocarpus moluccensis* in early June. Table 6 shows the phenogram of each mangrove tree during the period of our field trips.

Table 6: Phenogram of mangroves

Species	June	July	August	September
Avicennia marina	Vg	Vg	Vg	Fl
Bruguiera gymnorrhiza	En	En	En	En
Heritiera littoralis	En	En	En	En
Lumnitzera racemosa	Fl	Fl	Fl	Fl
Rhizophora mucronata	En	En	En	En
Sonneratia alba	En	Fr & Fl	Fr & Fl	Fr & Fl
Xylocarpus moluccencis	Fr &Df	En	En	En

Source: personal observations

Fr: fruit; Fl: flower; Vg: vegetative Df: defoliation

II.   3. Flora associated with mangroves

The families and species associated with mangroves are often found in the back of mangroves. They are represented in the table below. The species most associated with mangroves are: *Acrostichum, Guettarda, Pemphis, Ipomea, Euclea.*

Table 7: List of species associated with mangroves

Genres	Species	Families	Local name	Sites
Acacia	farnesina	MIMOSACEAE	Mguwu	Voidjou
Acrostichum	au^\^(^i^^	PTERIDACEAE	Mtsembea	Ikoni, Voidjou, Ouellah, Mohoro
Andasonia	digitata	MALVACEAE	Mbuwu	Ouellah, Domoimboini

Anacardium	*accidental*	ANACARDIACEAE	Mbibo	Mohoro
Caesalpinia	*bonduc*	FABACEAE	Mtso	Ikoni, Domoimboini, Voidjou,
Canthium	*bibracteatum*	RUBIACEAE	Mkararé	Mohoro
Dodonea	*viscosa*	SANPINDACEAE	Mtsouzimboi	Ikoni, Ouroveni, Seleani, Domoimboini
Euclea	sP	EBENACEAE	Mlala	Domoimboini, Seleani, Ikoni, Ouellah, Mohoro, Voidjou, Ouroveni
Flacourtia	*indica*	FLACOURTIACEAE	Mtsongoma ziba	Seleani, Ouroveni
Guettarda	*speciosa*	RUBIACEAE	Fulamboi	Ikoni, Voidjou, Ouellah, Domoimboini, Seleani
Hibiscus	*tiliaceus*	MALVACEAE	Mwaro	Ouroveni, Seleani, Ouellah, Domoimboini
Ipomea	*pes caprae*	CONVOLVULACEAE	Pumbu	Ikoni, Ouellah, Voidjou, Seleani, Ouroveni, Mohoro, Domoimboini
Lantana	*cammra*	VERBENACEAE	Trambamzugu	Seleani, Ouroveni, Voidjou
Morinda	*citrifolia*	RUBIACEAE	Mkonokono masera	Ikoni
Pandanus	sP	PANDANACEAE	Mtsango	Ikoni, Ouellah, Voidjou, Domoimboini,
Pemphis	*acidula*	LYTHRACEAE	Mdjana	Ikoni, Voidjou, Ouellah, Domoimboini
Scaevola	*sericea*	GOUDENIACEAE		Ikoni, Voidjou,
Scaevola	*plumieri*	GOUDENIACEAE		Seleani
Solanum	sP	SOLANACEAE		Ikoni, Ouroveni, Domoimboini
Tamarindus	*indica*	FABACEAE	Mhadju	Voidjou, Mohoro,

Source: personal data

III. ECOLOGICAL CHARACTERIZATIONS OF THE DIFFERENT TYPES OF MANGROVES

In our study areas, we distinguished two types of mangroves according to the morphology of the land.

* ❖ Mangrove of the lagoon type.
* ❖ Coastal mangrove.

Each type of mangrove presents different characterizations and physiognomies according to their morphology and their population. However, their floristic composition is almost similar. This characterization will be based on :

- the general physiognomy of the formation, in particular its height, density and cover;
- the floristic richness of the stands ;
- the nature of the surface soil.

The comparison of the two types of mangrove and their populations can explain the development of mangroves according to their station.

III.1. LAGOON-TYPE MANGROVE

A lagoon is a shallow body of water isolated from the sea by a barrier beach. Communication with the sea occurs through openings in the barrier beach.

The geomorphological aspect of the lagoon-type mangrove is very characteristic by its appearance interspersed with ancient barrier beaches. This type of mangrove is observed in 5 studied sites. The Rhizophoraceae family is omnipresent and largely dominant, in particular *Bruguiera gymnorrhiza*. But its physiognomy varies according to the morphology of the environment (nature of the substrate and presence of fresh water).

III.1.1- IKONI

This site is located in the west-south of Ngazidja in the region of Bambao at 4 km from Moroni on the RN2 road. The mangrove of Ikoni is located at 11°45.066 of latitude and 043°14.031 of longitude on an altitude varying between 2 to 24 m. It is a coastal forest with a large surface area subdivided into several lagoons and crossed by channels. It is separated from the sea by an ancient barrier beach, which has now been filled in to form a road. Developments (houses, a wedding hall, roads and a football pitch) form the boundaries of the mangrove. Of the species identified, *Bruguiera gymnorrhiza* is the most dominant. It is followed by *Lumnitzera racemosa, Sonneratia alba* and *Xylocarpus moluccensis*. The least represented species is *Heritiera littoralis*. (Photo 1).

Photo 1: Ikoni mangrove

a. *Zoning of the mangrove*

In this area, the following plant succession can be seen from the lagoon to the mainland:

- *Bruguiera gymnorrhiza* grouping in water and dune margins
- Mixed grouping of *Sonneratia alba,* and *Bruguiera gymnorrhiza*
- Mixed grouping of *Xylocarpu moluccensis* and *Lumnitzera racemosa*
- Behind the channel or in the back of the mangrove is an association of *Pandanus* sp, *Morinda citrifolia, Xylocarpus moluccencis* and *Heritiera littoralis*.

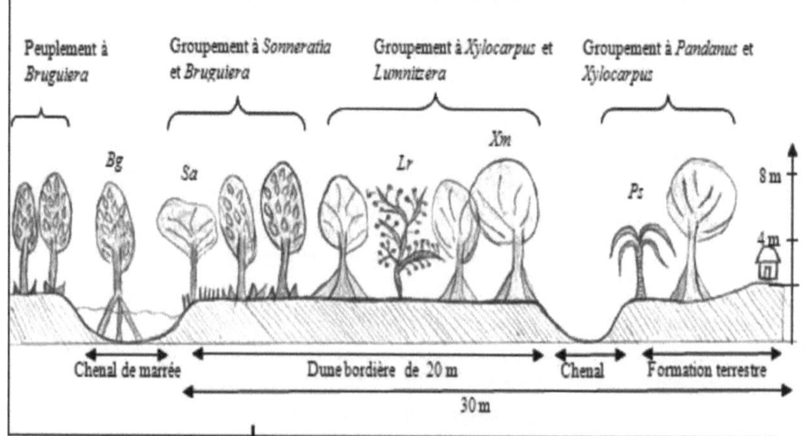

Figure 5: Schematic profile of the vegetation in Ikoni

Bg : Bruguiera gymnorrhiza; Sa : Sonneratia alba; Lr : Lumnitzera racemosa; Xm: Xylocarpus moluccensis; Ps: Pandanus sp

b. Vegetation structure

The stratification is not uniform. This mangrove has undergrowth plant formations and the height of each species was considered. Trees with relatively large trunks have an average diameter of 10 to 30 cm and are 3 to 7 m high. The most emergent species is *Xylocarpus moluccensis* with a maximum height of about 8 m and a maximum diameter of 50 cm.

The stand of *Bruguiera gymnorrhiza* occupies a very large area.
The edge of the channels is a place of sedimentation. *Lumnitzera racemosa* colonizes these areas first. By making a transect perpendicular to the channel, a stratification of the channel towards the interior of the vegetation is observed where *Bruguiera* is the dominant species.

c. Natural regeneration

In the Ikoni site, five species of mangroves were recorded. Only *Lumnitzera racemosa* and *Xylocarpus moluccensis* have good natural regeneration with regeneration rates of 484% and 142.846% respectively. The other species have a very weak regeneration. Therefore, the regeneration of this mangrove is average (table 8 and figure 4 Annex VI).

Table 8: Regeneration rate of mangrove species in Ikoni

Species	Regenerated individuals	Seed individuals	TR in %.
Bruguiera gymnorrhiza	19	24	79,17
Lumnitzera racemosa	121	25	484
Sonneratia alba	2	11	18,2
Xylocarpus moluccensis	40	28	142,86
Heritiera littoralis	1	0	0

Source: Personal data

d. Health status of the site

The Ikoni site shows good health, as vigorous individuals (with no axe marks: 24%) and normal individuals (with axe scars but still surviving: 74%) are very significant compared to individuals in poor condition (not surviving after cutting: 2%) (Figure 6).

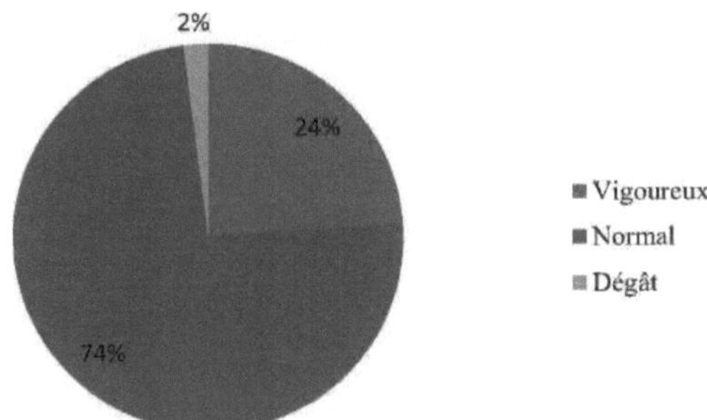

Figure 6: Health status of mangroves in the Ikoni site

III.1.2- VOIDJOU

This site is located in the west of Ngazidja in the region of Itsandra at 7 km from Moroni on the RN1. The mangrove of Voidjou is located at 11°38.229 of latitude and 043°15.845 of longitude on an altitude of 3 to 5 m. It is a littoral forest that was formed in a muddy area. It is separated from the sea by a barrier beach. It is limited to the West by the sea, to the East by the national road, to the North by the Military School and to the South by the National Television. (Photo 2)

Photo 2: Voidjou mangrove

a. Zoning of the mangrove

This area has four types of stands according to their floristic composition and structure. From the sea to the land, the structure of the formation is more developed.
- Mixed stand of nearshore and invasive species.
- Mixed stand of *Lumnitzera racemosa* and *Bruguiera gymnorrhiza.*
- Mixed stand of *Bruguiera gymnorrhiza* and *Avicennia marina.*
- Mixed stand of *Bruguiera gymnorrhiza, Lumnitzera racemosa, Acrostichum aureum.*
- On land, *Pandanus* sp and *Tamarindus indica* and other invasive species to land are found (Figure 7).

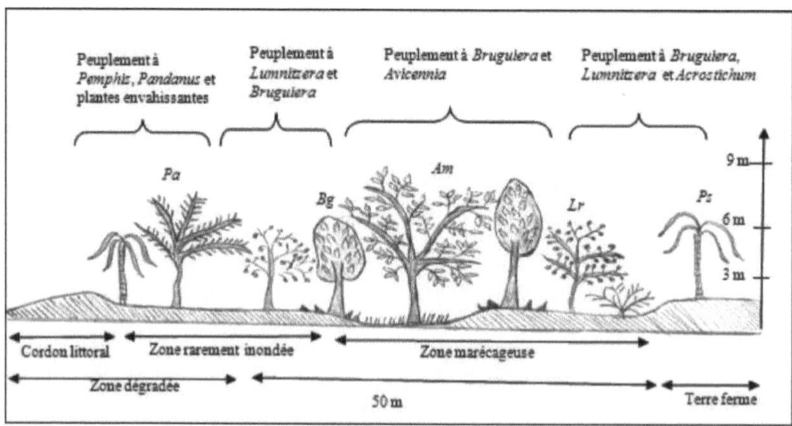

Figure 7: Schematic profile of Voidjou vegetation

Pa: Pemphis acidula; Bg: Bruguiera gymnorrhisa; Am: Avicennia marina; Lr : Lumnitzera racemosa; Ps: Pandanus sp

b. Vegetation structure

Behind the continuous cordon is a herbaceous and shrubby formation composed of *Pandanus* sp. *Ipomeapes caprae, Scaevola sericea, Guettarda speciosa, pemphis acidula...* and invasive plants. Stand 2 has a stratum dominated mainly by young *Bruguiera gymnorrhiza* plants accompanied by adult *Lumnitzera racemosa.* The height of the trees varies between 3 and 5 m and the diameter varies between 5 and 9 cm.

In the centre of the mangrove, there is a mixed stand of *Bruguiera gymnorrhiza* and *Avicennia marina* with a height varying between 3 and 9 m and a diameter between 10 and 84 cm. *Bruguiera* predominates on a muddy and stony soil full of waste.

The lower stratum from 0 to 2 m is the best represented. More than 300 seedlings easily fill a 5m2 area.

c. Natural regeneration

According to the inventory, this mangrove is made up of 3 species of mangroves including *Bruguiera gymnorrhiza, Lumnitzera racemosa, Avicennia marina. Bruguiera* is the species with a very good regeneration with a regeneration rate of 1340.75%. *Lumnitzera* and *Avicennia* have a good regeneration with respectively a regeneration rate of 750% and 143.75%. The natural regeneration of the mangrove is therefore very good (table 9 and figure 5 Annex VI).

Table 9: Natural regeneration rate of mangrove species in Voidjou.

Species	Regenerated individuals	Seed individuals	TR %.
Bruguiera gymnorrhiza	362	27	1340,75
Lumnitzera racemosa	30	4	750
Avicennia marina	23	16	143,75

d. Health status of the site

The health of this site is poor in that vigorous individuals 8% are in low numbers compared to normal individuals 75% and individuals in very poor condition have a rate of 17% (Figure 8).

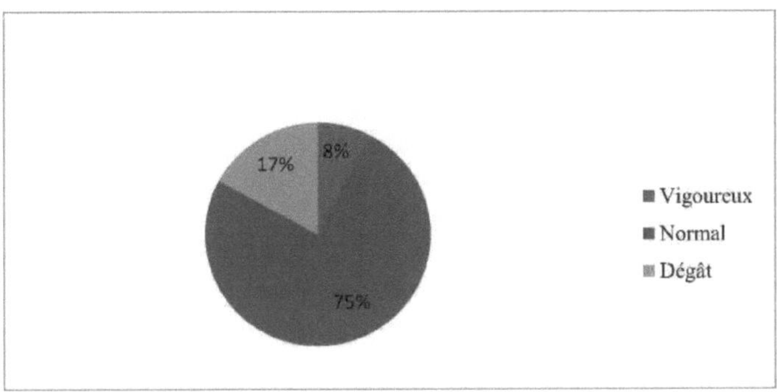

Figure 8: Health status of mangroves at the Voidjou site

III.1.3- SELEANI

This site is a lagoon mangrove, which is located in the east of the island in the region of Hamahamet at about 35 km from Moroni on the RN4. It is located at 11°30.625 of latitude and 043°24.145 of longitude approximately between 5 and 9 m of altitude. It has a large mangrove area compared to the other sites. It is separated from the sea by a very wide barrier beach of about 15 m. The substrate is very muddy so that a person not informed about the nature of the substrate could drown easily. The abundance of mosquitoes is an indicator of this. (Photo 3)

Photo 3: Mangrove of SELEANI

a. Zoning of the mangrove

In this site, the structure of the formation is highly developed. It is formed by a homogeneous stand dominated by *Rhizophora mucronata* in the centre mixed with *Lumnitzera racemosa*. The radius covered by the stilt roots is difficult to penetrate, although the species is still exploited. *Rhizophora* competes strongly with *Lumnitzera racemosa* because of its very dense and long network of stilt roots and its straight and tight trunks. The establishment of *Lumnitzera racemosa* in its rooting centre is almost impossible (Figure 11).

Overall, the Seleani mangrove is differentiated into four zones from the sea to the land.
- A degraded and mixed zone with *Euclea, Guettarda, Dodonea, Ipomea, Scaevola,*.
- A mixed zone with *Lumnitzera racemosa, Euclea* sp and *Hibiscus tiliaceus*
- A homogeneous zone with *Rhizophora mucronata* occupying a central position.
- A mixed zone of *Rhizophora mucronata* and *Lumnitzera racemosa* and *Acrosthicum.*

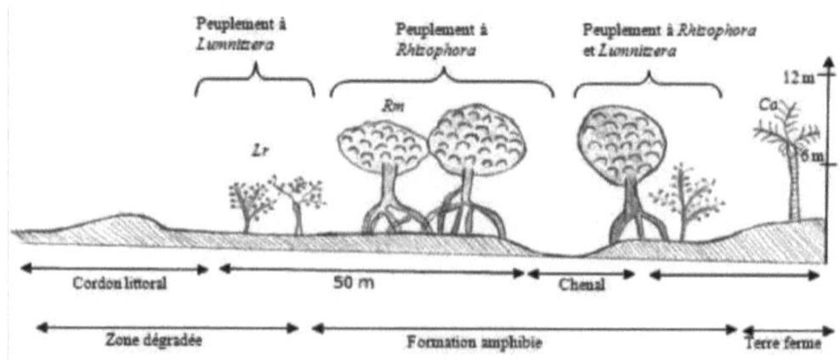

Figure 9 6: Schematic profile of the vegetation of Seleani

Lr: Lumnitzera racemosa; Rm: Rhizophora mucronata

b. Structure of the vertical

Behind the barrier beach is a stand of herbaceous plants and shrubs. It is a degraded environment, where individuals are widely spaced. Most of the species are companion plants to the mangrove and their size does not exceed 2 m. The dominant species are *Euclea* sp and *Dodonea viscosa.*

The *Lumnitzera racemosa* stand is a dense shrub layer, 2 to 6 m high, on soil that is rarely flooded. This species easily fills one unit every square meter, but in a few places it is associated with *Euclea* sp. The upper stratum is totally *Rhizophora mucronata*, whose tops range from 7 to 12 m.

c. Natural regeneration

Two stands predominate in this site: the *Rhizophora mucronata* stand and the *Lumnitzera racemosa* stand. Only the latter has good regeneration with a rate of 119.23% while *Rhizophora* regeneration is weak with a rate of 57.14%. Thus, the natural regeneration of this mangrove remains weak (Table 10 and figure 6 in Annex VI).

Table 10: Regeneration rate of mangrove species in Seleani

Species	Regenerated individuals	Seed individuals	TR %.
Rhizophora mucronata	12	21	57,14
Lumnitzera racemosa	31	26	119,23

d. Health status

The Seleani mangrove has a very good state of health in that vigorous and normal individuals have rates of 46% and 44% respectively while individuals in poor condition are only 10% (Figure 10).

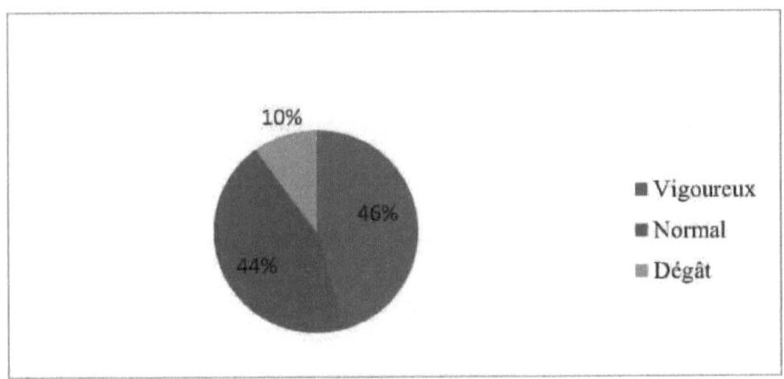

Figure 10: Health status of mangroves in the Seleani site

m.1.4 - DOMOIMBOINI

This site is a lagoon mangrove, it is located in the North-West of Ngazidja in the region of Mboudé at about 25 km from Moroni on the RN1. It is located at 11°29 .250 of latitude and 043°16.577 of longitude on an altitude varying between 3 to 4 m. Among the identified species, *Bruguiera gymnorrhiza* has a very high abundance-dominance. Then come *Lumnitzera racemosa, Rhizophora mucronata,* and *Sonneratia alba.* The latter is the least abundant species represented by four plants (photo 4).

Photo 4: Mangrove of Domoimboini

a. *Zoning of the mangrove*

This mangrove is characterized from the sea to the land by discontinuous stands interspersed with large rocks:

- A stand of *Bruguiera gymnorrhiza*
- A mixed stand of *Bruguiera gymnorrhiza, Rhizophora mucronata*
- A mixed stand of *Pemphis acidula, Lumnitzera racemosa*
- A mixed stand of abundant *Pemphis acidula* and *Euclea* sp

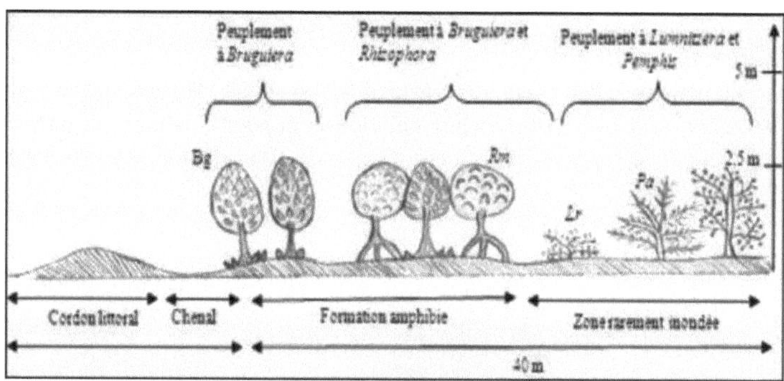

Figure 11: Schematic profile of the vegetation of Domoimboini

Bg : Bruguiera gymnorrhiza; Rm : Rhizophora mucronata; Lr : Lumnitsera racemosa; Pa: Pemphis acidula

b. *Vegetation structure*

The height of the trees varies according to the substrate. In the southern part of the mangrove, the substrate is sandy and the trees are up to 5 m high. Whereas, in the northern part, the substrate is very rocky and the trees are very stunted and not very developed. In this northern part, there are two strata:

❖ Stratum from 0 to 2 m dominated by *Lumnitzera* and growing most often in rocks.

❖ Stratum of 1 to 3 m occupied mainly by *Bruguiera* accompanied by *Rhizophora. Pemphis acidula is a* bushy species with a wide distribution. It develops as a woody liana and its tops do not exceed 2 m.

<u>***NB***</u>: The small size and stunting of the individuals are due to the rocky substratum. The existence of monospecific series of *Bruguiera* and *Rhizophora* as well as the stunting of the mangroves means

that the individuals are spaced out and do not manage to retain the light and therefore the cover is nil.

c. Natural regeneration

This mangrove is made up of 4 species of mangroves including *Bruguiera, Lumnitzera, Rhizophora,* and *Sonneratia. Bruguiera* and *Lumnitzera* are the species with a good regeneration with respective regeneration rates of 175.75% and 173.33%. *Rhizophora* and *Sonneratia* have poor regeneration with regeneration rates of 36.84% and 0% respectively. The natural regeneration of the mangrove is therefore average (tableaullet (figure 7 Annex VI).

Table 11: Regeneration rate of mangrove species in Domoimboini

Species	Regenerated individuals	Seed individuals	TR %.
Rhizophora mucronata	7	19	36,84
Lumnitzera racemosa	26	15	173,33
Bruguiera gymnorrhiza	58	33	175,75
Sonneratia alba	0	1	0

d. Health status of the site

The state of health of the site is very good with vigorous individuals having a rate of 80%, while normal individuals have a rate of 15% and dead individuals have a rate of only 5% (Figure 12).

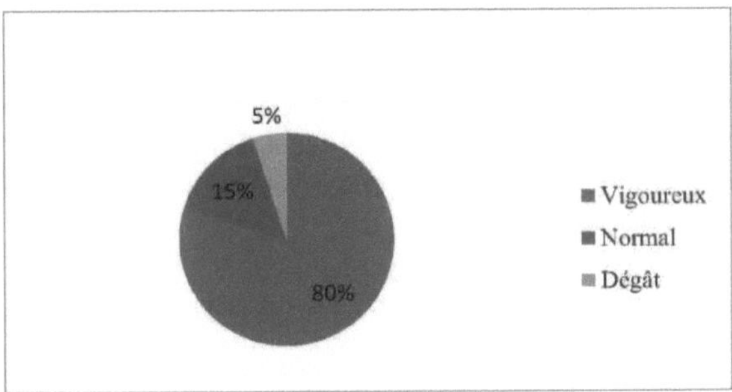

Figure 12: Health status of mangroves in the Domoimboini site
III.1.7- MOHORO

The mangrove of Mohoro is of the lagoon type. It is located in the South-East of the island in the region of Badjini at about 5 km from Foumbouni (third big city of the island) on the RN3. It is located at 11°49.388 of latitude and 43o28.002 of longitude with an altitude of 3 to 6 m. It is separated from the sea by a barrier beach. It is constituted by a single species of mangrove *(Bruguiera gymnorrhiza)* accompanied by some associated plants. This mangrove has a very low density with only 8 seedlings and a few seedlings. Its natural regeneration is very weak and its health is disastrous. Some individuals are in decomposition, so we can conclude that this mangrove is in a state of regressive evolution. (Photo 5)

Photo 5: Mohoro mangrove

Health status of the site

The state of health of the site is generally poor, insofar as it has been noted that there used to be a dense mangrove with several species of mangroves. But with the threats (deforestation, grazing and cultivation in the vicinity of the mangrove, cleaning fires) that weigh on this site, only *Bruguiera* manages to survive and adapt with difficulty to the environmental conditions.

III.2- COASTAL MANGROVE

III.2.1- OUROVENI

A littoral mangrove has an orientation parallel to the coast. It was observed at Ouroveni. This site is located in the south of the island in the Badjini area, about 6 km from Fombouni on the RN3. It is located at 11°54 .756 of latitude and 43°30.039 of longitude approximately between 3 to 7 m of altitude. The substratum is very muddy. It is separated from the sea by a wide, sharp tanne of about 15 m. Most of the tanne is devoid of vegetation, and it is periodically flooded during spring tides. The vegetation of the tanne herbaceous is a discontinuous formation of herbaceous plants. (Photo 6).

Photo 6: Mangrove of Ouroveni

a. Zoning of the mangrove

This mangrove presents four types of settlement according to their floristic composition and their structure. From the sea to the land we find :
- Monospecific stand with *Sonneratia alba* ;
- Stand with *Rhizophora mucronata* and *Sonneratia alba* ;
- Mixed formation with *Euclea* sp, *Dodonea viscosa*, *Ipomea pes caprae*, *Solanum* sp,

Flarcoutia indica localized in the edge of the tanne;
- Stand with *Lumnitzera racemosa* and *Hibiscus tiliaceus* located behind the tanne.

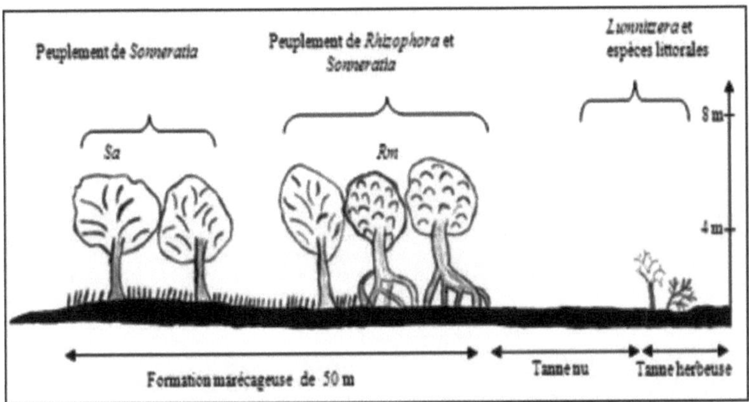

Figure 13: Schematic profile of the vegetation of the Ouroveni mangrove

Sa: Sonneratia alba; Rm: Rhizophora mucronata

b. Vegetation structure

Tree heights vary in all three stands.

❖ The stand of monostratified *Sonneratia alba* is observed on immature (non-compact) clay soil. The trees with relatively large trunks (15 to 40 cm in diameter) are 4 to 7 m high.

❖ The stand of *Rhizophora mucronata* occupies a very large area. This species is sometimes accompanied by *Sonneratia alba*. This stand, like that of *Sonneratia, does* not present any stratification apart from the seedlings. Its height varies between 4 and 6 m and the diameter can reach 36 cm.

❖ The edge of the tanne is an area colonized by a discontinuous formation of herbaceous plants and shrubs. This formation is characterized by associated species and invasive species.

❖ Finally, at the very back of the mangrove is a mixture of *Lumnitzera racemosa* and *Hibiscus tiliaceus* occupying a solid environment that is rarely flooded. The height of *Lumnitzera* varies between 2 and 4 m, while that of *Hibiscus* can go up to 6 m.

c. Natural regeneration

This site is represented by three mangrove species including *Sonneratia alba*, *Rhizophora mucronata* and *Lumnitzera racemosa*. The respective regeneration rates of *Rhizophora* and *Sonneratia* are 41% and 20.68%, while that of *Lumnitzera* is 138.88%. Thus the natural regeneration of this mangrove is poor (table 12, figure 8 Annex VI).

Table 12: Regeneration rates of mangrove species in the Ouroveni site

Species	Regenerated individuals	Seed individuals	TR %.
Rhizophora mucronata	22	53	41
Sonneratia alba	6	29	20,68
Lumnitzera racemosa	25	18	138,88

Source: personal data

d. Health status of the site

The health status of the Ouroveni mangrove is normal as most mangroves are normal with a rate of 85%. While the vigorous individuals have a rate of 10% and the dead individuals are only 5% (Figure 14).

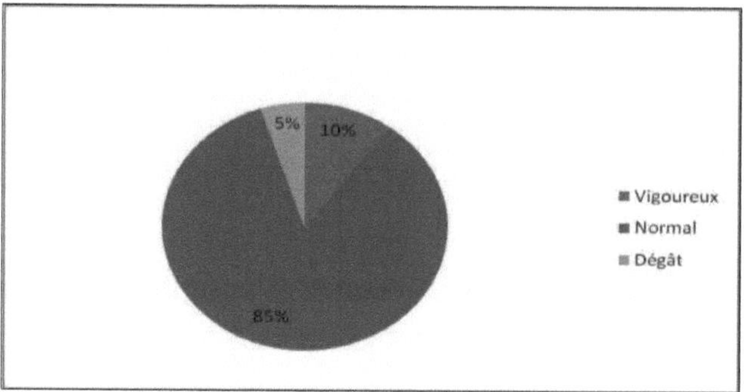

Figure 14: Health status of mangroves in the Ouroveni site

III.2.2- OUELLAH

This site, which is a coastal mangrove, is located in the north of Ngazidja Island, about 7 km from Mitsamihouli (the island's first major town) on the RN3. It is strongly degraded, and the ground is very sandy. It is characterized by 5 species of mangroves including *Bruguiera gymnorrhiza, Avicennia marina, Xylocarpus moluccensis, Lumnitzera racemosa, Sonneratia alba.* These last two species are represented by a few feet. (Photo 7)

Photo 7: Ouellah mangrove

a. Zoning of the mangrove

This mangrove is characterized from the sea to the land, by :
* Stand with dominant *Bruguiera gymnorrhiza* and *Pemphis acidula;*
* Mixed stand of *Avicennia marina* and *Bruguiera gymnorrhiza* ;
* Mixed stand of *Avicennia, Xylocarpus,* and *Bruguiera;*
* Stand with *Xylocarpus moluccensis,* and *Bruguiera gymnorrhiza* ;

These stands are in a degraded state because of their massive exploitation and the silting to which they are exposed (Figure 15).

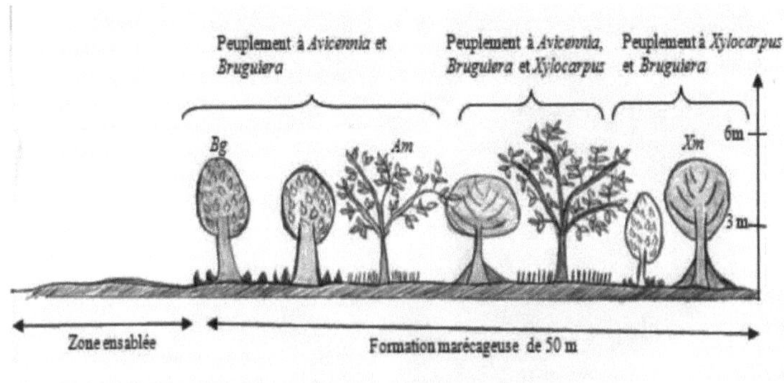

Figure 15: Schematic profile of Ouellah vegetation
Bg: Bruguiera gymnorrhiza; Am: Avicennia marina; Xm: Xylocarpus moluccensis

b. Vegetation structure

The plant formation of this mangrove does not present a uniform stratification. In the North-Western part, the trees easily reach 10 m while in the South-East the formation is discontinuous and the trees do not exceed 6 m in length. *Bruguiera is the* most abundant species in our study area.

c. Natural regeneration

The regeneration rate of *Bruguiera* is 70.6 %, that of *Avicennia* is 36 %, on the other hand that of *Xylocarpus* is null. The natural regeneration of this mangrove is thus weak. (Table 13 and figure 9 Annex VI).

Table 13: Regeneration rate of mangrove species in Ouellah

Species	Regenerated individuals	Seed individuals	TR %.
Bruguiera gymnorrhiza	24	34	70,6 %
Avicennia marina	9	25	36 %
Xylocarpus moluccensis	0	13	0

Source: personal data

d. Health status of the site

The state of health of the site is very poor, as individuals in very poor condition are significant at 52%, while normal and vigorous individuals are 37% and 11% respectively, figure 16.

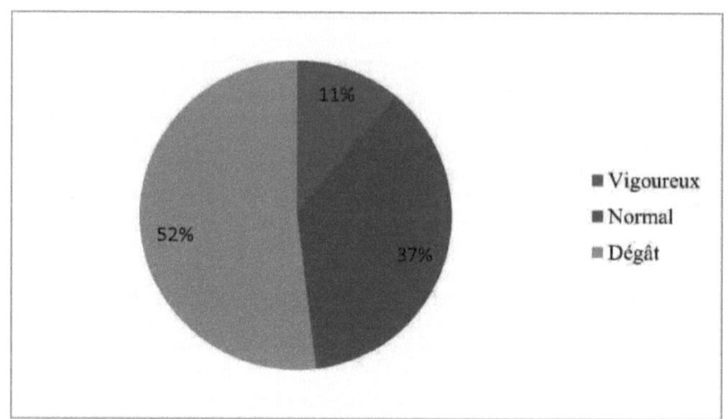

Figure 16: Health status of mangroves in the Ouellah site

IV.   DISTRIBUTION OF MANGROVES IN NGAZIDJA

The cartographic method made it possible to highlight the distribution of the studied mangroves in the whole island. From this parameter, we could know the phytogeography of the target mangroves. Map 5 is used to locate the study sites and to provide information on the distribution of all the mangroves on the island of Ngazidja.

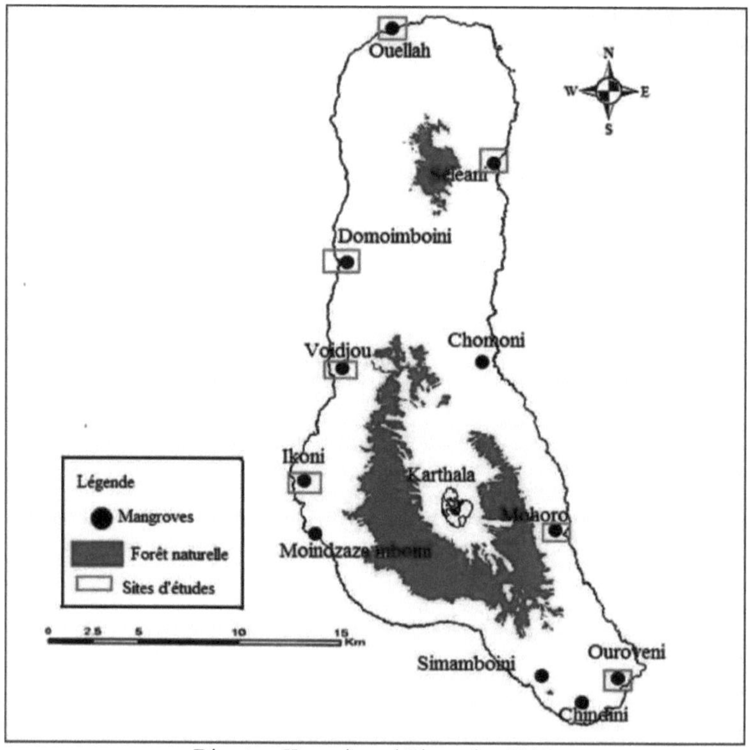

Director: Kamaria and El-yamine 2012
Map 5: Current distribution of mangroves in Ngazidja

<h1 style="text-align:center">V.   BIOMASS RATE AND CARBON STOCK</h1>
<h2 style="text-align:center">V.   1. Estimation of carbon stock at each site</h2>

Forests play a direct role in the cycle of the main greenhouse gas, carbon dioxide. The ability of mangroves to develop in difficult environments with particular ecological and phyto-sociological conditions is the reason for their high plant productivity, which reflects their great capacity to fix atmospheric carbon through photosynthesis.

The method applied by FAO consists in determining the carbon stored in the different study sites. The calculation is based on the measurement of DBH >10 cm and the maximum height of mangroves. From these measurements we were able to establish a summary table on biomass and carbon stocks for each site. The carbon stock result is expressed in Kg/ha. The unit of surface considered is 100 ^{m2} equivalent to 0.01 ha per site.

Table 14: Biomass and carbon stock rates at the different sites

Sites	Maximum height of a mangrove (m)	No. of plants/0.01ha	Z (DBH) in cm	Biomass in kg/ha $BA = 0.544. [Z (DBH)]^{2}. Ht$	Carbon sequestered per site in Kg/ha $C = BA \times 0.5$
IKONI	8	49	700,5	21355,274	10677,637
VOIDJOU	9	20	420,5	8657,119	4328,559
DOMOIMBOINI	2,5	56	810	8922,96	4461,48
OUROVENI	8	28	529	12178,68	6089,34
SELEANI	12	19	380,5	9451,254	4725,627
OUELLAH	6	22	338	3728,924	1864,462
TOTAL					32147,105

Source: personal data

<h2 style="text-align:center">V.   2. Assessment of the capacity of each mangrove to sequester carbon</h2>

Table 14 shows the average carbon stock in mangrove biomass per unit area per site. Some sites sequester more carbon than others. The analysis of the results shows that the dry biomass and the quantity of carbon vary according to the density and the maturity or age of the trees. These results show that:

❖ The Ikoni site has a biomass quantity of 21355.274 kg/ha, which undoubtedly leads to a sequestration of atmospheric carbon also three times higher than the others. The analysis shows that this important quantity evolves with the density of the trees as well as with maturity (diameter and height). This justifies their capacity to sequester carbon: the carbon stock is 10677.637kg/ha or 10.677 t/ha.

❖ The analysis of carbon stocks in Ouroveni and Seleani sites shows that their biomass rates depend on age or maturity (very high diameter and height). Since the trunk is the main carbon reservoir and large trunks and trees sequester much more carbon than small ones. These sites sequester 6089.34 kg/ha or 6.089 t/ha and 4725.627 kg/ha or 4.725 t/ha respectively.

❖ In the Domoimboini site, the analysis shows that the small size of the mangroves reduces the biomass. However, the biomass is less low because of the density of mangroves (56 feet/0.01ha). This site has a carbon stock of 4461.48 kg/ha or 4.461 t/ha.

❖ The Voidjou site has a very low density with a value of (20 feet/0.01ha), but most individuals also have a DBH < 10 cm. However, the mangroves considered in this analysis are mature with DBHs

49

up to 55 cm. Hence, this site has an average carbon stock of 4328.559 kg/ha or 4.328 t/ha.

❖ In the Ouellah site, we found that deforestation reduces the frequency of individuals. This phenomenon accompanied by the small size of mangroves considerably reduces their biomass and consequently this site sequesters less carbon with a stock of 1864.462kg/ha or 1.864 t/ha.

Using this method, it was possible to show that the mangroves of Ngazidja are capable of accumulating a maximum of more than 10 tonnes of carbon per hectare, with variations depending on the general characteristics of the site.

_**NB**_: Some sites have a large surface area, notably Seleani, Ikoni, and Ouroveni, which considerably increases their capacity to sequester carbon.

By comparing our results with those obtained from the study of *estimation of the quantity of carbon stored by the reforested mangrove of the Saloum delta in Senegal,* we found that the mangrove of Ngazidja sequesters a significant quantity of carbon with a stock of 32147.105 kg/ha of total carbon that is to say 32.147 t/ha in all the inventoried sites. However, as the area of the island is small, that of mangroves is certainly small so we can argue that carbon sequestration is low compared to other mangroves previously studied in other countries.

It is important to note that these results are difficult to compare with other results obtained in other studies due to the lack of data on the exact area of mangroves in Ngazidja.

VI.   FAUNAL DIVERSITY

The mangrove of Ngazidja, is characterized by an exceptional fauna, abundant, but poor in species. It is difficult to characterize its real fauna because some animals can be seen occasionally called "visitors", some have their habitat permanently and others are "associated". Most of the species found in this environment are adapted to the harsh conditions. The majority of birds and reptiles of the mangroves often come from neighbouring environments and stay there to feed or nest.

VI.   1. INVERTEBRES

The invertebrates of the mangrove are much more represented by insects, crustaceans and molluscs.
   ❖ Insects abound, especially mosquitoes;
   ❖ Molluscs (gastropods and oysters) such as *Terebralia palustris, Buccinides* sp and *Haupinites* sp which attach themselves abundantly to the stilt roots and pneumatophores of mangroves;
   ❖ Crustaceans are very numerous, especially crabs and shrimps.

V.   2. VERTEBRES

VI.   2.1. Fish
Some fish, because of their mobility and easy change of environment, frequently enter freshwater and mangrove areas. Among the most frequent species, we distinguish :
   ❖ *Periophtalmus koelrenteri* (jumping fish) is a characteristic species of the mangrove (UNEP, 2002);
   ❖ *Anguilla* sp.
VI.2.2. Reptiles

Reptiles are not very numerous in the mangroves, among the frequent species are lizards

(Table 15).

Table 15: Some lizards frequenting mangroves

Scientific names	Families	Vernacular names
Phelsuma v-nigra comoraegrandensis	GECKONIDAE	Ignandroboe
Trachylepis comorensis	SCINCIDAE	Maboya / Nguzi
Hemidactylus platycephalus	SCINCIDAE	Nkafiri

Source: personal observations

VI.2.3. Birds

Some birds can associate with mangroves. They are called associated animals. These are mainly marine birds (table 16).

Table 16: Some very common birds in mangroves

Scientific name	Family	Vernacular name	Location
Egretta dimorpha	ARDEIDEAE	Mgweda (Egret dimorpha)	Ouroveni
Butorides striatus rhizophore	ARDEIDEAE	Mnadjidi (Green Heron)	Ouroveni
Ardea alba	ARDEIDEAE	Ngnamandé (Great egret)	Ikoni, Ouroveni
Casmerodius albus melarhynchos	ARDEIDEAE		
Bubulcus ibis	ARDEIDEAE	Djamoimbe (Ox heron)	Ikoni, Ouroveni
Phaeton lepturus	PHAETONITIDAE	Nkahorindi	Ouroveni
Numenius arquata		Soulouloupvu	Ouroveni

Source: (Awardine 2012) + surveys and personal observations

Some forest and non-forest birds are threatened and find refuge in mangroves and spend time there or find their ecological niche (food, tranquillity, and nesting). This is the case of the Ouellah mangrove where several birds have taken refuge (table 17).

Table 17: Some birds nesting in the Ouellah mangrove

Scientific names	Family	Vernacular names
Streptopelia picturata	COLUMBIDAE	Bwantsi
Cinnyris humbloti humbloti		Ntsihintsi
Hypsipetes madagascariensis	PYCNONOTIDAE	Sopve manga
Zoosterops maderaspatana kirki	ZOSTEROPIDAE	Gnandronga
Corvus albus	CORVIDAE	Gawa

Source: Personal observations

a) *Coenobita* sp sur un tronc de *Bruguiera*

b) *Uca urvillei* sur le tanne

c) *Cardisoma* sp sur le tanne

d) *Trachylepis* sur un tronc de *Xylocarpus*

e) *Phelsuma v nigra comoraegrandensis* sur les racines de *Rhizophora*

f) *Terebralia palustris* entourant un pied de *Lumnitzera* à marée basse

g) Oiseau sur un palétuvier

h) Groupe de *Bubulcus ibis* sur les cimes de *Bruguiera*

i) *Ardea alba* chassant sur le tanne

j) Nid d'oiseau sur un *Bruguiera*

k) Petit poisson capturé dans la mangrove

l) *Peronia* sp dans la boue de mangrove

Plate 4: Fauna of the Ngazidja mangrove

VII. PRESSURES AND THREATS TO MANGROVES

Surveys carried out among local populations and observations made in the field have made it possible to identify anthropogenic and natural threats to mangrove habitats and mangroves.

VI. 1. Anthropogenic threats

These threats are mainly caused by the misuse of different resources from the mangrove ecosystem. The main ones are:

VII. 1.1- Cutting of mangrove wood

Mangrove wood is mostly exploited locally for fuel and construction. The cutting of mangroves, which is the most degrading local anthropic factor of the mangrove, can indeed disturb the dynamics and the renewal of mangroves.

VII. 1.2- Traditional fishing and exploitation of crabs and shrimps

Some villagers practice traditional fishing and use toxic products *(Tephrosia)* and materials (dynamites) that are harmful to the mangrove ecosystem. The collection of crabs and shrimps in an abusive way also becomes disastrous because these animals play a role on the ecology of mangroves.

VII.1.3- Overexploitation of sand in coastal areas

Some villagers irrationally remove sand from the beaches and coastal areas, which results in the retreat of the beaches and causes adverse effects on the beachfront mangroves.

VII.1.4- Animal husbandry and roaming around the mangrove

Cattle rearing on the land edges at low tide is one of the negative factors of the mangrove. Zebu roaming hinders natural regeneration, as these animals graze on young seedlings.

VII.1.5- Fires and Clearing

Some mangrove areas are deforested and set on fire in order to transform them into cultivation fields or make them accessible and hospitable. This is the case of the mangrove of Mohoro which is in a state of regressive evolution.

VII.1.6- Coastal occupation and construction around mangroves

As the population increases, the need for habitable and usable land increases. This is the case in some villages where the mangrove area is used for building houses, filling in roads and placing football pitches. All these activities contribute to the imbalance of this ecosystem.

VII.1.7- Garbage dumping in the mangrove

Some communities consider the mangrove as a dumping ground for household waste, plastics, pollutants and all other forms of waste. All these wastes can modify the physico-chemical properties of the soil and cause the death of mangrove trees.

VII.1.8- Fish farming

The installation of fish ponds in the Comoros is done in lagoons. As observed during the surveys, these installations have been carried out by the authorities of Ngazidja Island. However, the fish introduced and bred in these ponds come from South Africa and Sudan, which constitutes a threat for the indigenous fish and shrimp species.

a) Coupe de *Lumnitzera racemosa*

b) Décharge d'ordure dans la mangrove

c) Pâturage aux alentours de la mangrove

d) Extraction de sable dans les plages

e) Coupe rase des palétuviers

f) Construction aux alentours de la mangrove

Plate 5: Anthropogenic pressures and threats
VII.2 Hypothetical dynamics of the mangrove

Observations show that the mangrove is under the influence of several anthropic factors responsible for degradation. This indicates a regressive evolution. Figure 17 shows the degradation process of the mangrove ecosystem

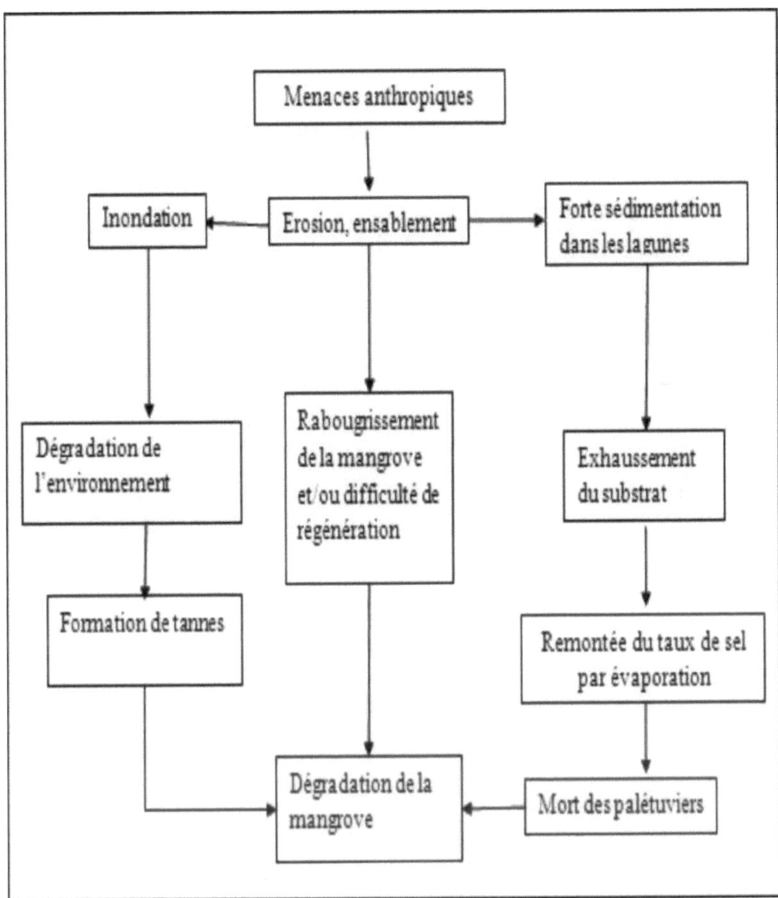

Figure 17: Hypothetical diagram of mangrove dynamics

PART 4:
DISCUSSION AND RECOMMENDATIONS

DISCUSSION AND RECOMMENDATIONS

I. DISCUSSION

1.1. Problems encountered in the field

The working methodology of this study and the results obtained allow us to discuss some points.

1.1.1. Methodological issues

The readings were taken at low tide which is from 7am to 5pm. However it happens sometimes that the high tides persist until 10 am or that the water starts to rise from 3 pm. So the readings were taken following the rise and fall of the tides. The structure of the ground constitutes a major constraint for the realization of the statements, insofar as certain environments are: either difficult to access, or discontinuous, or not homogeneous.

1.1.2. Problems with investigations

The work of the surveys required a lot of flexibility and understanding of the respondents, because the inhabitants are suspicious and give vague answers. Some people refused to answer. Most of the time we did the survey in groups to better approach the subject. There are other constraints because most of the respondents do not know the vernacular names of mangroves. For them, all the plants found in the mangrove are called "Mhonko" but are used in various ways. Sometimes it is only through the answers that one understands which species it is.

1.2. Floristic results

The floristic procession includes only the species met in the studied sites. More than 30 species distributed in 23 families were recorded in the mangroves. However, the floristic analysis shows that the mangrove of Ngazidja is floristically rich but shows the absence of endemic species.

The density of mangroves for all the sites studied varies from 19 to 56 feet per 100 ^{m2}. Thus, large trees are less numerous. Of all the mangroves found on the sites studied, only *Bruguiera gymnorrhiza* and *Lumnitzera racemosa* have good regeneration. On the other hand, in Ndzouani it is *Rhizophora mucronata* that has a good regeneration (Boura, 2009).

Moreover, invasive species such as *Cissus quadrangularis, Lantana camara, Acacia farnesia,* were encountered. The presence of these species and their low density shows that this ecosystem is in a degradation phase.

NB: According to the remarks drawn on the zonation, density, floristic results and physiognomy state of these mangroves, this ecosystem does not benefit from good ecological conditions allowing its good development. Special attention should therefore be given to the mangrove.

I.3. Exploitation of mangroves
1.3.1- Illegal exploitation

In the past, the exploitation of mangrove timber did not pose a threat to the stands, but this situation has been reversed over time. The cutting of mangroves, which remains local, especially for firewood and construction, is becoming increasingly important and can threaten the stability of these ecosystems. Mangrove clearing remains the most degrading local anthropogenic factor of Ngazidja mangroves, as the woody products taken from the mangrove represent an important quantity for the fuel supply of the inhabitants living near the mangroves.

1.3.2- Use of mangroves

The valorisation of mangrove resources in medicinal plants is not really exploited. It is true that the leaves, roots and fruits of mangroves are used in traditional medicine in Ngazidja but the therapeutic powers of these species are not yet known by everyone; only a few people have knowledge of the virtues of these plants. Their pharmaceutical value seems to be effective but they need to be studied and verified.

In Ngazidja, some of the uses of mangroves are similar to those in other countries, notably in Madagascar or in Black Africa. *Rhizophora* is used as gaulettes and poles in the construction of houses, *Avicennia* and *Xylocarpus* are used to make pirogues. The barks of Rhizophoraceae are used as dyes for traditional articles. On the other hand, the manufacture of charcoal from mangrove trees does not exist in Ngazidja, unlike in other countries. The sale of mangroves is not a common practice in Ngazidja. The mangrove hinterland is not used for salt making or rice cultivation like in Madagascar. So we think that the conservation of the mangrove of Ngazidja is possible.

1.3.3- Use of fishery resources

The exploitation of the fishery resources of the mangrove of Ngazidja is limited to the fishing of fish and rarely to shrimps. The mangrove is not used for shrimp and crab fishing in Ngazidja, because Comorians are not used to eating shellfish. However, it is a very developed activity in other countries like Madagascar. According to the surveys carried out, the small shrimps found in the mangrove of Ikoni have almost disappeared following the installation of a fish farm in the lagoon. According to the same source, some foreigners collect crabs in the Voidjou mangrove. Is it not necessary to reinforce the law in this area?

I.4. Natural threats and impact of climate change on mangroves

Natural threats and global warming have a major impact on the physico-chemical properties of mangroves. (Roger, 2007). These properties are influenced from the following abiotic factors:

1.4.1- Coastal silting

Coastal silting is a natural phenomenon that occurs continuously due to waves, winds and sand deposits from the land during the rainy season. These sand deposits promote the formation of beaches and the silting of certain areas of the coastal strip. This sandy strip is increasingly advancing towards the open sea and the mangroves. As a result, the vegetation of some mangroves is in a critical state (Roger, 2007).

1.4.2- Temperature

The rise in temperature can cause the capillary rise of the water table, the water evaporates and the salt remains. The salinity level increases at the mangrove soil level. Therefore, some mangrove trees may succumb to this phenomenon, but this threat is not felt directly (Roger, 2007).

1.4.3- Precipitation

The decrease or sudden increase in rainfall is felt in the mangroves. Heavy rainfall can increase the amount of water infiltration and accentuate the phenomenon of erosion. The sediments accompanying the erosion directly invade the mangroves and constitute a danger for the health of this ecosystem. It should also be noted that a significant decrease in rainfall can indirectly cause the mangroves to dry out and the mangroves to die (Roger, 2007).

1.4.4- Sea level

The permanent sea level can increase from its usual level, due to the phenomena of ice melting and the abundance of precipitation at the time of heavy rains. (Roger, 2007). This causes flooding in the mangroves.

1.4.5- Greenhouse gases (GHGs)

Five sectors have been identified as sources of GHG emissions in Comoros. These sectors include: land use change and forestry, agriculture, energy, waste, solvent use, etc. They play a key role in the condensation of air masses in the atmosphere. This phenomenon causes an irregularity of precipitations and consequently an over-salting at the level of the marsh, and a drying of the mangroves. The mangroves are asphyxiated due to intense evapotranspiration and the channel water dries up. By classifying the emission sources according to their importance, we notice that the land use change and forestry sector represents the main source with 775,524 tons of CO_2, followed by agriculture with 459,957 tons of CO_2, then energy with 70,524 tons of CO_2 and waste with 9,963 tons of CO_2 (DGE, 2002).

I.4.6 Invasive plants

Some plants that invade the back mangroves can even advance to the mangroves and compete with the mangroves, or in the extreme case, they can live as parasites on the mangroves and hinder the dynamics of the mangroves. This is the case of *Cissus quadrangularis, Acacia farnesina, Lantana camara, Flacourtia indica.*

I.5. Carbon sequestration by mangroves

Mangroves are also one of the best natural ways to protect the planet from global warming, because they have a high carbon sequestration capacity. This characteristic of mangroves urgently requires preventive measures. One of the most important contributions offered by mangroves is their ability to sequester atmospheric carbon and store it in the biomass and swamp substrate. According to the February 2007 issue of National Geographic, "mangroves are carbon factories.

The massive loss of coastal wetlands has a huge and unseen impact with the oxidation and release of carbon stored in mangroves. Through research by Dr. Ong of Sams University of Malaysia, it was learned that the soil layers that make up the mangrove substrate contain a large volume of carbon that reaches 10% or more. Each hectare of mangrove sediment can contain nearly 700 tonnes of carbon per metre of depth. Clearing mangroves and excavating their substrate could result in the oxidation of 1400 tonnes of carbon per hectare per year.

According to Dr. Ong, "If we assume that only half of this volume is oxidized over a 10-year period, we would have 70 tonnes of carbon returning to the atmosphere per hectare per year for ten years. That's about 50 times the sequestration rate. Thus, only 2% of mangroves need to be converted for all the benefits of mangroves as atmospheric carbon sinks to be lost. (www.ideecasmance.org).

According to the FAO, the current rate of mangrove loss is 1% per year, i.e. about 150,000 ha of mangroves are lost each year. This translates into a loss of trapping capacity of about 225,000 tonnes per year, plus the release of about 11 million tonnes of carbon that was stored in the mangrove soil.

Clearly, this is a huge problem that requires concerted action. Not only are we losing the important carbon sequestration capacity of mangroves, but we are also seeing the release of large amounts of polluting gases from their substrate. The continued removal of mangroves, for whatever reason, must be seen in a different light.

In our study we found that mangroves in Ngazidja also have a high capacity to store carbon. This capacity is better appreciated at the level of the sites with large surface like Ikoni, Ouroveni, and Seleani. However, we also observed a strong degradation of the mangrove due to anthropic activities. The current rate of mangrove degradation is seriously threatening these fragile ecosystems and decreasing their ability to mitigate the effects of climate change. Hence the importance of a management and conservation policy for the mangrove of Ngazidja.

I.6. Current interests of the mangroves of Ngazidja
I.6.1 Socio-economic interests of mangroves

Mangrove forests, once considered as "wastelands", are now recognized as an important ecosystem mainly because of their socio-economic interests. In our country where the demographic pressure on coastal areas is high, mangroves are becoming natural living resources of paramount importance. They are traditionally used by local populations for a large number of purposes. In addition, it is recognized that they provide tangible benefits including: (timber, firewood or charcoal, construction wood, medicinal and medicinal plants, fisheries resources ...).

I.6.2. Ecological interests of mangroves

Mangroves play several functions in the marine and coastal ecosystem. They provide indirect benefits such as

a. Protection of coastal areas

Mangroves remain one of the best natural means of protecting marine and coastal ecosystems. They are well equipped to adapt to changes in prevailing winds and seas and to rising sea levels. They protect shorelines by smothering wave and wind energy, regulating the quantity of nearshore water through their stilt roots and pneumatophores, sedimentation and retention of soil nutrients (Roger E, 2007).

b. Water purification and protection of aquatic ecosystems

Mangroves and mangrove soils play a crucial role in water purification. Mangroves protect corals and underwater seagrass beds by forming biological filters through the retention and export of sediments and nutrients and by preventing siltation by absorbing pollutants from urban effluents and agricultural activities upstream. They thus prevent the downstream process of eutrophication, which can promote the rapid growth of plants and algae and lead to a depletion of oxygen levels in the environment (Jeannoda, 2008).

c. Animal refuge and nursery area

The mangroves of Ngazidja offer a real spawning area for many fauna species. They are resting, security, nesting and reproduction environments for some animals. They provide the necessary nutrients for fish, crabs and shrimps. Some species spend part of their time in mangroves or are part of the food chain linked to this ecosystem. The degradation of mangroves will have negative repercussions on the survival of this biodiversity (personal work).

II. RECOMMENDATIONS

II.1 Recommendation for mangrove conservation

The protection and rehabilitation of mangroves is one of the major concerns of the world. The management of natural resources in countries with galloping demography is confronted with the unfavourable anthropic factor. Ignorance and low income of the riparian populations, associated with the involvement of several structures in the use of resources, accelerate the degradation of the ecosystem, thus the complexity of mangroves requires a participatory management.

The role of each partner, the orientation of socio-economic activities, the application of legislation, the awareness and education of the populations concerned, the initiation of research programmes, the place of NGOs, etc. are all points that must be defined in order to determine effective strategies for reliable results. A sustainable management policy must be organized around a multidisciplinary team to direct the various activities towards this ecosystem.

Table 18: Logical framework for the implementation of a mangrove management and conservation system

	Intervention logic	Objectively verifiable indicator (OVI)	Sources of verification
Overall objective	Ecological characterization of mangroves, and assessment of carbon stock by mangroves		
Specific objectives	Ecological and biological study of mangroves and identification of pressures and threats to mangroves	Biological inventory of each site; Mapping of surveyed sites.	Final Report
	Identification of socio-economic and ecological interests	Ethnobotanical and socio-economic surveys	Final Report
	Suggestion of a management and conservation system for this ecosystem		Final Report
	Estimation of the quantity of carbon stored by mangroves	Carbon stock results	Final Report
Expected results	• Biodiversity of the inventoried mangrove ; • Biological and physico-chemical characteristics of the mangrove identified ; • Impact of anthropic activities on mangrove assessed ; • Population resource needs assessed, reserve security for preservation assessed ;	Results and interpretations	

	• Areas of use, conservation, and rehabilitation determined; by setting operating criteria and standards; • Populations sensitized to the rational management and conservation of mangroves ; • Forest carbon sink developed by the growth and reforestation of mangrove areas; • Strategies for the use and preservation of these defined resources; • classification of mangroves as part of the proposed protected areas.			
Activities	• Bibliographic research ; • Ethnobotanical surveys ; • Ecological surveys ; • Quantification of stored carbon ; • Mapping of the different mangroves of Ngazidja ; • Activities for the conservation of mangroves : **a.** Establishment of a mangrove zoning: delimitation of the (use, conservation and rehabilitation zone); **b.** Establishment of a waste management system and prohibition of dumping of garbage in mangroves; **c.** Creation of IGAs (improved agricultural and fishing practices; training of eco-tourist guides); **d.** Development of ecotourism in these areas; **e.** Continued awareness raising and training of villagers.			

Source: Personal data

<u>**NB**</u>: International cooperation will need to be involved through the UN organizations responsible for environmental protection and development. Organizations such as the World Bank, the World Conservation Union (IUCN), the World Wide Fund for Nature (WWF), and the International Society for Mangrove Ecosystems (ISME) can be called upon to bring their full weight to bear and influence governmental decisions towards management.

11.2. Integrated Management Recommendation

Through the development of integrated management, the law should be strict for the sake of the environment. But some accompanying measures should therefore be implemented to remedy the economic problems and to improve the living conditions of the villagers following the ban on resource exploitation.

Alternative financial resources should then be developed. Some proposals that are feasible in Ngazidja can be mentioned:
❖ To develop agriculture which until now has only known archaic sectors and traditional practices;
❖ Develop fishing activities (fish, crabs, shrimps)
❖ Develop beekeeping (improve honey production)
❖ Develop fish preservation techniques (drying, freezing, salting).

11.3. Recommendation for mangrove reforestation

The reforestation of mangroves is essential to slow down continental erosion, preserve the fauna it shelters and ensure carbon sequestration. This requires a certain amount of training for the villagers concerning the techniques to be adopted, the planting method and the care to be taken.
It is recommended to choose the species for reforestation according to the characteristics of the site. To have good results on a reforestation in Ngazidja, the following species are recommended:

❖ *Avicennia marina* and/or *Sonneratia alba,* for the oversalted areas in the coastal mangroves (Ouroveni and Ouellah) flooded by the tides.
❖ *Rhizophora mucronata* and/or *Ceriops tagal* in areas flooded by high tides, with relatively high salinity, such as lagoons;
❖ *Bruguiera gymnorrhiza* and/or *Lumnitzera racemosa* for areas close to the mainland, rarely flooded, with lower salinity than the species mentioned above.

II.4.Integration of a policy for the implementation of key sites

The choice of key sites will be based on the vulnerability of these mangroves to anthropic actions. Among the sites studied, some are still in good condition (Ikoni, Ouroveni), others have already undergone withdrawals (Seleani), if not overexploitation (Ouellah, Voidjou). The sites can be classified into three categories:

a) Key sites for ecological restoration: some mangroves are under very intense pressure. The ecological function may no longer be ensured. In this case, restoration is needed.

b) Key sites for rehabilitation and enrichment: ecological functions are still maintained but pioneer species enrichment is needed.

c) Key sites for conservation: Some mangroves are still intact. They need to be kept in good condition and conserved.

Table 19: Summary of key sites

Sites	Important features	Risks and threats	Alternatives
Ikoni	Type lagoon, mangrove still intact, rich biodiversity, beautiful landscape, large area.	City planning, Construction timber	c)

Voidjou	Lagoon type, degraded and diminished mangrove.	Overexploitation of woods with heavy waste disposal in the vicinity	a)
Domoimboini	Lagoon type, degraded and diminished mangrove, possibility of regeneration, beautiful landscape.	Silting Construction timber Waste disposal	b)
Mohoro	Lagoon type, possibility of regeneration	Grazing in the mangrove forest Clearing fires Traditional culture	a)
Seleani	Lagoon type, large area, mangrove still intact	Sand mining Mangrove grazing Fuelwood	c)
Ouroveni	Coastal type, rich biodiversity, site protected by the village community	Sand extraction Herd migration Fishing, firewood	b) & c)
Ouellah	Coastal type, rich biodiversity, degraded mangrove, beautiful landscape	Silting Overexploitation of wood Sand extraction	a)

II.5. Traditional mangrove conservation

Some regions consider mangroves as sacred places, others consider them as places haunted by evil spirits which makes some people afraid to venture into the mangrove. This situation plays in favour of the protection of this ecosystem.

CONCLUSION

CONCLUSION

Mangroves, which are part of the floristic and faunal wealth of the Comoros, play an important role in the natural heritage of these islands, given the special habitat and structural originalities of these natural heritages. They represent one of the most productive forests in terms of biomass but also the most threatened because of their fragility. Indeed, their conditions of adaptation are very difficult and their exploitation or management requires great attention. Mangroves have particular functions, notably ecological, socio-economic and climate change mitigation functions.

During this study we were able to provide some basic knowledge on the ecology and current state of the mangrove of Ngazidja in the face of anthropization, climate change and knowledge on the evaluation of carbon stock. Thus, the results of the field work gave ample information on mangrove species, more precisely their ecology, distribution, use and capacity to sequester carbon.

Seven species of mangroves were particularly observed in the field, namely *Rhizophora mucronata, Avicennia marina, Sonneratia alba, Bruguiera gymnorrhiza, Lumnitzera racemosa, Xylocarpus moluccensis and Heritiera littoralis,* which are divided into six families: Rhizophoraceae, Avicenniaceae, Sonneratiaceae, Combrétaceae, Malvaceae and Meliaceae. The Rhizophoraceae family is the most represented in Ngazidja. However, the species *Ceriops tagal* and *Xylocarpus granatum do* not exist in Ngazidja but rather in the other islands.

Two types of mangrove are found in Ngazidja: the coastal type and the lagoon type. The vegetation in these two types of mangrove is generally dense but encounters some difficulties on the natural regeneration and the state of health.

In spite of the pressures and threats that the mangroves of Ngazidja undergo, these sites fully fill their carbon sink by sequestering in spite of the difficulties, more than 10 tons of carbon per ha in Ikoni, and more than 4 tons of carbon per ha in the other sites. And for all the sites studied, we found an average of 32.147 tons per ha. This high carbon sequestration shows a good contribution of mangroves in the mitigation of climate problems. However, efforts remain to be made in terms of scientific research of other mangroves.

However, the absence of a management and protection system and the lack of legislative texts concerning their conservation aggravate their degradation. It is in this context that we invite all competent authorities of Comoros and organizations working for the protection of the environment to take into account the real threats in order to act accordingly, and to promote an exchange of knowledge and experience in the field of conservation and management of mangrove resources in Comoros.

Moreover, in terms of research, our study is very limited by the time factor and the lack of technical and financial resources. Thus, it is far from being complete. For this reason, we recommend further work of the same kind, especially on the determination of the surface area of mangroves and the carbon stock of other mangroves not studied.

Finally, this work constitutes a basic tool for future researchers who want to pursue studies in this framework, but also a basic tool that can contribute to the rational management of mangroves and the mitigation of climate change.

BIBLIOGRAPHY

BIBLIOGRAPHY

I- BIBLIOGRAPHICAL REFERENCES

ABDILAHI, M. M. 2009. *Etude de la végétation à baobab des îles Comores (flore, écologie et dynamisme) : cas de Mohéli et de la Grande-Comores.* Dissertation of DEA in plant biology and ecology, University of Antananarivo. 81p.

ADJANOHOUN, J; Aké, A; AHMED, A. 1982. *Contribution aux études ethnobotaniques et floristiques aux Comores* A.C.C.T Paris 215p.

AHAMADA. 2005. *Study of the role of local communities in the sustainable management of coastal resources: the case of Comoros.* Master II, Marne Vallée University, France. 82p.

AMANN, C; AMANN, G; ARHEL, K; GUIST, V; MARQUET, G. 2011. *Plants of Mayotte*

ANDILYAT, M. A. 2007. *Ecological study of the Mount Karthala forest (Grande-Comores): Ethnobotany, Typology, Natural regeneration, Spatiotemporal evolution and potential zonation in conservation site.* DEA thesis in plant biology and ecology, University of Antananarivo, 86p.

ANDRIAMALALA, C. 2007. *Ecological study for mangrove management in Madagascar: Comparison of a littoral and estuary mangrove using remote sensing.* Thesis. Antananarivo, 280p.

ANLLAOUDINE, A. H. 2009. *Ecological characterization of the most used plant species and their habitats on the northern part of the grid massif (Grande-Comores): (inventory, ethnobotany, ecology and mapping).* DEA thesis of biology and plant ecology, University of Antananarivo 75p.

ASMA-ILHOUSNA, M. 2012. *Ecological and descriptive approach of the genus Tambourissa of the family MONIMIACEAE in Ngazidja and Ndzouani (Comoros).* Master II thesis in sustainable development and biodiversity conservation. University of Comoros. 84p.

AWARDINE, M. S. 2012. *Avian phenology and geographical distribution of migratory birds wintering in Grande-Comores*: International Master II, University of Comoros 84p.

BACAR, D. G. 2010. *Characterization of Rhizophora* spp. *plantations and attempt to quantify the carbon sequestered by reforestation: case of Darssilamé sérère village. (Senegal).* Master II in forestry and environment, University of Thiès Senegal. 54p.

BATTISTINI, R and VERIN, P. 1984. *Geography of the Comoros.* Agence de Coopération Culturelle et Technique, Ed Nathan, Paris. 142p.

BEAUDOIN, G. 2003. *The Third Way for the Kyoto Protocol: Carbon Sequestration. General consultation on the implementation of the Kyoto Protocol in Quebec,* Department of Geology and Geological Engineering, Faculty of Science and Engineering, Laval University, Quebec. 15p.

BOUDOURESQUE, F. C. 2009. *Structure and functioning of marine benthic ecosystems. The mangrove ecosystem.* Master of Oceanography, Centre d'Océanographie Marseille. 68p.

BOURA, M. 2009. *Ecological and ethnobotanical consideration of mangroves and its mangrove trees in Anjouan Island for conservation* (cartography, ecology, *development* scheme, *threats and pressures and management plan).* Dissertation DESS en Gestion de Ressources Naturelles et Environnement, Université de Toamasina 86p.

CADET, Th. 1980. *The vegetation of Reunion Island: Phytoecological and phytosociological study,* Doctoral thesis, University of Aix-Marseille 3, France. 312p.

CHEKHIDINE, S. M. 2007. *Ethnological and ethnobotanical consideration and conservation of plants around the RAMSAR site "Dziani Bunduni" (Djando Mohéli Comores).* Dissertation DESS en Gestion de Ressources Naturelles et Environnement, University of Toamasina 98p. Commission de l'Océan Indien. 2011. Study of vulnerability to climate change, qualitative assessment (Comoros). 114p.

CONSERVATION COMOROS. 2005. *Biodiversity and resource use assessment and environmental awareness.* Preliminary report. CNDRS/ACA/ AID. 20p.

DAROUSSI, A. 2006. *Etude écologique des espèces les plus utilisées dans l'île d'Anjouan : Cas du mont Ntringui* ; Mémoire de DEA, Ecologie Végétale. Université d'Antananarivo. 92p. DECELLE, J. E. 1980. L'enthomofaune comorienne. Africa-Tervueren. 87p.

DEGUE-NAMBONA, R. M. 2007. *Contribution of mangrove reforestation in the Saloum delta (Senegal) to atmospheric carbon sequestration: Case of Djirnda and Sanghako villages.* Dissertation for the DEA in environmental sciences, Cheikh Anta Diop University, Senegal. 100p.

DIN, N and BLASCO, F. 2003. *Sustainable management of mangroves under demographic pressure and impoverishment. 445p.*

DIRECTION GENERALE DE L'ENVIRONNEMENT. 1993. *Diagnosis of the state of the environment in Comoros.* UNDP, UNESCO/IUCN project.

DIRECTION GENERALE DE L'ENVIRONNEMENT. 2001. *Rapport national sur l'environnement marin et côtier aux Comores. 39p.*

DIRECTORATE GENERAL FOR THE ENVIRONMENT. 2002. *United Nations Framework Convention on Climate Change.* 1st National Communication 56p.

DIRECTION GÉNÉRALE DE L'ENVIRONNEMENT. 2007. 2nd National Communication: *Climate Change Consultation: Analysis of Greenhouse Gas Mitigation Options.* 21p.

DODANE, C. 2007. *Wood, forests and carbon in France: climate and energy issues and prospects.* Geo-confluences web edition on sustainable development and geographical approaches, University of Lyon, UMR / CNRS 5600 Environnement Ville et Société. ENS LSH.

DONALD, J; MACINTOSH; ELISABETH, C; ASHTON. 2005. *Principles for a code of conduct for the sustainable management and use of mangrove ecosystems.* World Bank, ISME, CenTER Aarhus. 89p.

DUPONEY, J. L; PIGNARD, G; HAMZA, N. 2006. *Carbon sequestration in forests. UMR-écologie et écophysiologie forestières*. INRA Nancy France. 71p.

DUVIGNEAUD, P. 1980. *The synthesis of ecology.* 2nd edition France.

EPHYSE, L. D. 2009. *Carbon offsetting and the Aquitaine wood sector.* INRA. 30p.

FAO. 2005. Global Forest Resources Assessment 2005: *Thematic study on mangroves in the Comoros - country profile.* FAO Forestry Department, Rome. Preliminary study. 23p.

FAIDATI, D. 2007. *State of the current situation of the mangrove around Tuléar.* DEA de biologie végétale Université de Toliara 82p.

FOUAD, A. R. 2010. *Assessment of mangroves (Comoros).* UNDP/COSEP. 45p.

GUILLET, M; RENOUX, E; ROBIN, M; DEBAINE, F; RAKOTONAVALONA, H. D; RATSIVALAKA, S. 2008. *Monitoring and analysis of the evolution of the Mahajamba mangrove (Northwestern Madagascar).* Report of the international multidisciplinary symposium. 8p.

JEANNODA, V and ROGER, E. 2008. *Collection of articles on mangroves of Madagascar. Department of Plant Biology and Ecology,* HONKO, University of Antananarivo.250p.

KEITH, P; ABDOU, A; LABAT, J. N. 2006. *Faunal inventory of the rivers of the Comoros and botanical inventory.* Muséum national d'histoire naturelle. 64p.

LAITAT, E; PERRIN, D; SHERIDAN, M; LEBEGUE, C; PISSART, G. 2004. *A model for the calculation of carbon sequestration by forests, according to the terms of the Marrakech Accords and Belgium's reporting commitments to the Kyoto Protocol.* EFOBEL Plant Biology Unit. University of Gembloux (Belgium). 40p.

LANCE, K; KRENYIEN, C and RAYMOND, I. 1994. Extraction of forest products of parc and buffer zone and long term monitoring. Report to Park Delimitation Unit, WCS/PCDIM, Tananarive.

LEBIGRE, J. M. 1983. *Les tannes, approches géographiques.* Centre Universitaire Régional de Tuléar. 63p.

LEBIGRE, J. M. 1990. *Les marais maritimes de Gabon et Madagascar: contribution géographique à l'étude d'un milieu naturel tropical.* Thesis for the doctorate of state Institute of Geography University of Bordeaux III. 673p.

LOUETTE, M; MEIRITE, D; JOQUE, R. 2004. *The terrestrial fauna of the Comoros Archipelago.*

MAOULANA, A. S. 2010. *Geobotanical diagnosis and anthropic impacts on the mangrove landscape in Toliara Bay.* Multidisciplinary doctoral thesis, geography option, University of Toliara. 76p.

MAUMONT, S; BOUSQUET-MELOU, A; FOUGERE-MANEZAN, M. 2002. *Phylogeny and bio-geological history of mangroves.* Dossiers gestion forestière/mangroves, laboratoire d'écologie terrestre, Université Paul-Sabatier. 52p.

MBAE, A; HAMADI. 1991. *Protection de la faune endémique des Comores.* Dissertation, ENES, Mvouni. 51p.

MELIUS, A. 2003. *Radar measurements of the dynamics of Guyanese mangroves.* DEA thesis, Physical Methods in Remote Sensing, University of Paris 7. 18p.

MICHEL, S. 1995. *Ecological approach and dynamics of the mangrove ecosystem in the Morondava region.* Dissertation of DEA in plant ecology, University of Antananarivo. 65p. MINISTRY OF RURAL DEVELOPMENT, FISHERIES, HANDICRAFTS AND THE ENVIRONMENT. 2006. *National Action Programme for Adaptation to Climate Change* (PANA). UNEP/GEF. 77p.

MOULY, A. 2009. *Systematic study of the Rubiaceae of Mayotte and Comoros.* Rapport final d'expertise. 156p.

MSAIDIE, D. M. 2005. *Indication for the participatory assessment of vulnerability and adaptation to climate change (Grande-Comore coastal zone)* Final report. 21p.

OLIVIER, B. 1998. *The mangrove in Madagascar: a natural wealth to be managed.* ORSTOM actualités, Paris.

UNEP and DGIC, 2002. *East Africa: Atlas of Coastal Resources in the Comoros.* RFIC. 154p.

RAMADE, F. 2008. *Encyclopedic dictionary of natural sciences and biodiversity,* Dunod. Paris.

RAMADHOINI, A. I. 2011. *Floristic biodiversity of the Dibwani plateau in Grande-Comore (caves and savannahs), determination key for grasses.* Dissertation DESS-SE Biologie de Conservation, University of Antananarivo. 62p.

RICHMOND, M. D. 2002. *A field guide to the seashores of eastern Africa. And the western Indian Ocean Islands.* Department of Biology and Environmental Science, University of Kalmar, Sweden and coordinator of the Marine Science Program of Sida (1989 - 1999).414p.

ROGER, E. 2007. *State of the mangrove of Madagascar with regard to climate change.* Report, University of Antananarivo. 30p.

ROGER, E; RAJERIARISON, C; RAKOUTH, B. 2007. *Collection of documents for the ecological monitoring of the environmental program.* TOHIRAVINA 2 483p.

ROTH, P. L. 1964. *Natural regeneration in tropical forests. Dipterocarpus drey (Dau) on the Cambodian side of the Gulf of Siam. Bois des tropiques Madagascar Comores,* Dissertation, ENES, Mvouni. 38p

SCHOENE, D. 2002. *Assessing and reporting changes in forest carbon stocks.* 65p.

TELLO, R. 2008. *The employment effects of the development of CO2 capture and storage in France, a macroeconomic study compared with renewable energies.* Master in economic analysis and risk governance, University of Versailles. 79p.

TOUNG, D. 2010. *Estimation of the quantity of carbon stored by a forest in reconstitution: Case of a young fallow in the classified forest of Mondah.* Ingénieur des Techniques des Eaux et Forêts, Ecole Nationale des Eaux et Forêts du Cap-Estérias (Gabon). 82p. YAHAYA, I. 2007. *Contribution à l'inventaire et l'étude de la flore indigène du massif du Karthala.* Master's thesis, Université Marne la vallée. 151p.

II- WEBOGRAPHIES

www.iepf.org/ressources/livre drills.../Forets Chap.../page46.html *(03-11-2012)*

www.fao.org/docrep/013/al480F/al480F.pdf *(30-11-2012)*

www.unep.org/.../Comoros National State of Coast Report.pdf *(03-11-2012)*

www.equatorinitiative.org/images/stories/.../mdg 1348163266.pdf *(03-11-2012)*

http://wrmbulletin.wordpress.com/osd.xml*(03-11-2012)*

http://www.fao.org/WAICENT/AGICULT/agll/globdir/index.htn (18-10-2012)

http://www.africanbirdclub.oSrg/countries/comores (05 - 10- 2012) www.ideacasamance.org(20-09- 2012) www.actioncarbone.org(05 - 10- 2012) http://www.memoireonline.com(21- 11-2012) www.ecologie.gouv.fr/IMG/pdf/final (30-10-2012) http://www.unesco.org/en/tentativelits/5107 (20- 09- 2012) http://www.malango-comores.com/histoire-peuplement.htm(20- 09- 2012)

ANNEXES

ANNEXES

ANNEX I: CLIMATE VARIATIONS

Table: Monthly variations of rainfall (mm) and temperature (C°) from 1971-2000 and 2011

	Jan.	Feb.	March	April	May	June	Jul.	August	Seven.	Oct.	Nov.	Dec.
Monthly rainfall 2011 in mm	363.9	770.2	373.7	249.1	128.5	331.3	134.7	125.7	179.8	132.8	71.9	178.1
Normal rainfall 1971-2000	341.0	249.4	256.3	266.8	209.0	264.4	181.0	123.2	84.6	112.2	114.3	199.7
Average temperature 2011 in C°.	28.1	27.9	27.9	28.7	27.3	26.1	25.1	24.5	25.2	26.5	28.1	27.5
Normal temperature 1971-2000	27.6	27.9	27.8	27.4	26.4	25.1	24.2	23.9	24.4	25.6	26.7	27.5

Source: Moroni weather station, ANACM 2012

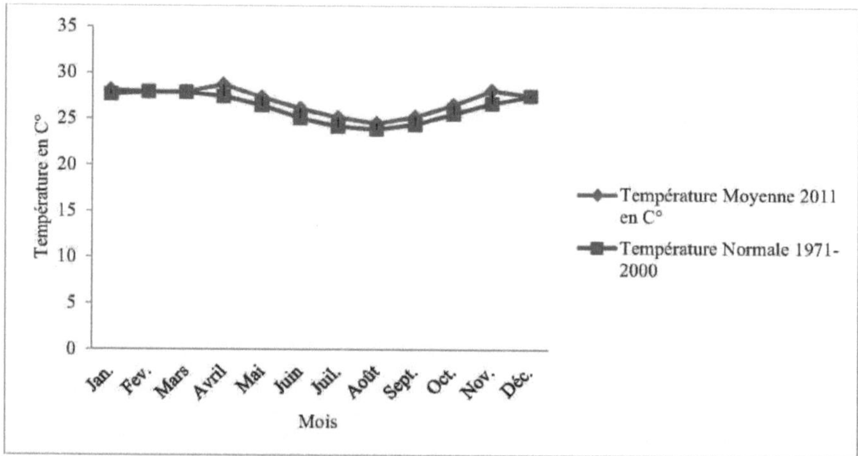

Figure 1: Monthly temperature curve for 1971-2000 and 2011 (Source: Moroni weather station, ANACM 2012)

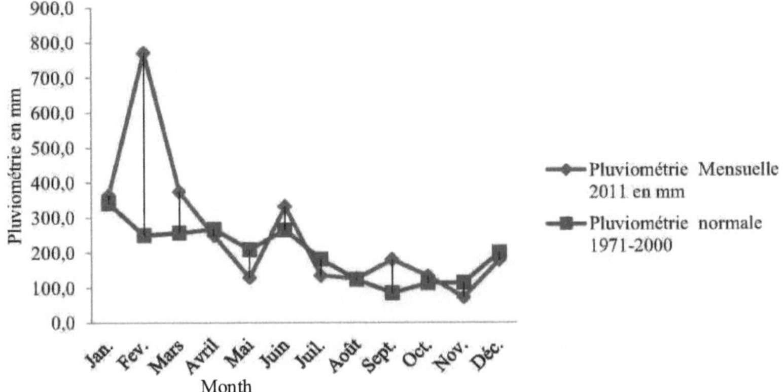

Figure 2: Rainfall curve for 1971-2000 and 2011 (Source: Moroni weather station, ANACEM 2012)

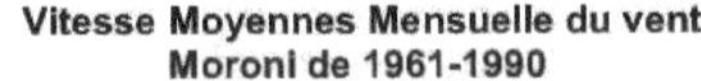

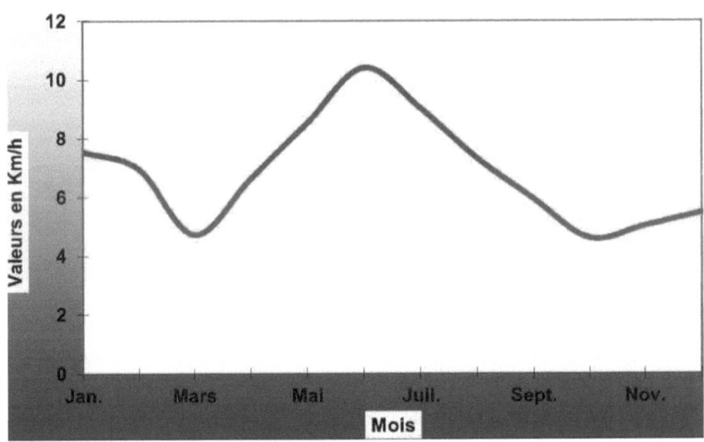

Figure 3: Monthly wind speed Moroni station (1971- 2000) (Source: Moroni meteorological station: ANACM, 2012)

APPENDIX II: ETHNOBOTANICAL SURVEY FORM

1. **Identification of the respondent**
   > Date
   > Village
   > Name and surname of the respondent
   > Sex and age
2. **Main use of the mangrove**

   > What does the mangrove represent for you?
   > What are the local traditions related to the mangrove?
   > What are the activities of the mangrove (before and now)?
   > What are the use patterns of mangroves and associated species?
     a. Firewood
     b. Construction timber
     c. Lumber
     d. Fishing
     e. Harvesting of medicinal plants
     f. Other
3. **Target species**
   > What are the target species?
   > What is the vernacular name of each species?
   > What is the use of each species?
   > What is the method of preparation of medicinal plants
   > What part of the plant is used?
     - Sheets
     - Flowers
     - Fruits
     - Stems
     - Bark
     - Root
4. **Fauna**
   > What are the animal species found in the mangrove?
   > What is the local name for each species?
5. **Mangrove management**
   a) Is there a system for protecting the mangrove?
   > Established by the local community
   > Established by the State
   b) Have you ever seen a mangrove reforestation?

6. **Evolution and history of the mangrove**
   > Have you noticed any changes in biodiversity in the mangrove?
   > What are the causes of its changes?
   > Has accessibility in the mangrove changed since you were there?

APPENDIX III: ECOLOGICAL SURVEY FORM

Ecological Inventory Sheet: TRANSECT

Name
Date
Location
Altitude
GPS coordinates
Survey area
Stratum of belonging

Plot number	Scientific names	Dendrometric parameters			Phenology			Health		
		DHP (cm)	HF (m)	HT (m)	Fl	En	Vg	V	N	D
Plot 1	Esp1									
	Esp2									
Plot 2										
Plot 3										
Plot 4										

Fr: fruiting
Fl: flowering
Vg: vegetative
V: vigorous
N: normal
D: damage

Ecological Inventory Sheet: PLACEAU

Name: Location:
Date: Survey area :
Altitude : Orientation :
GPS coordinates: Plot area :

phenology	Health (V,N,D)
height	DHP

Scientific name	Number of individuals	Plot 1		Plot 2		Plot 3		andte 4	
Esp1									
Esp2									
Esp3									

Fr: fruiting

Fl: flowering

Vg: vegetative

V: vigorous N: normal

D: damage

APPENDIX IV : TABLE : GLOBAL FLORISTIC MAP FOR ALL STUDIED SITES

Families	Genres	species	Vernacular names
ANACARDIACEAE	*Anacardium*	*accidental*	Mbibo
AVICENNIACEAE	*Avicennia*	*marina*	Mhonkomewu
COMBRETACEAE	*Lumnitzera*	*racemoza*	Mhonkomché
CONVOLVULACEAE	*Ipomea*	*Pescaprae*	Pumpu
EBENACEAE	*Euclea*	*sp*	Mlala
FABACEAE	*Caesalpinia*	*bonduc*	Mtso
	Tamarindus	*indica*	Mhadju
	Leucaena	*leucocephala*	
FLACOURTIACEAE	*Flarcourtia*	*indica*	Mtsongoma ziba
GOUDENIACEAE	*Scaevola*	*sevicea*	
	Scaevola	*plumieri*	
GUSIACEAE / GUTTIFERAE	*Calophyllum*	*inophyllum*	Mkorwa
LYTHRACEAE	*Pemphis*	*acidula*	Mdjana
LECYTHIDACEAE	*Barringtonia*	*asiatica*	
MALVACEAE	*Hibiscus*	*tiliceus*	Mwaro
	Heritiera	*littoralis*	Mhonko
	Andasonia	*digitata*	Mbuwu
MELIACEAE	*Xylocarpus*	*moluccensis*	Mfubu
MIMOSACEAE	*Acacia*	*farnesia*	Mguwu
PANDANACEAE	*Pandanus*	*sp*	Mtsango
PTERIDACEAE	*Acrostichum*	*aureum*	Mtsembea
RHIZOPHORACEAE	*Bruguiera*	*gymnorrhiza*	Mhonkoshuma
	Rhizophora	*mucronata*	Mhonkommé
RUBIACEAE	*Guettarda*	*speciosa*	Foulamboi
	Morinda	*citrifolia*	Mhonkomasera
	Canthium	*bibracteatum*	Mkararé
SAPINDACEAE	*Dodonea*	*viscosa*	Mtsouzimboi
SOLANACEAE	*Solanum*	*sp*	
SONNERATIACEAE	*Sonneratia*	*alba*	Mhonkomouigni
VERBENACEAE	*Lantana*	*camara*	Trambamzungu
VITACEAE	*Cissus*	*quadrangularis*	Dungawungwe

Photo 1 : Pneumatophores de *Sonneratia alba*

Photo 2 : Pneumatophores d'*Avicennia marina*

Photo 3 : Racines échasses de *Rhizophora mucronata*

Photo 4 : *Pemphis acidula* développant des racines échasses pour s'adapter aux conditions du milieu

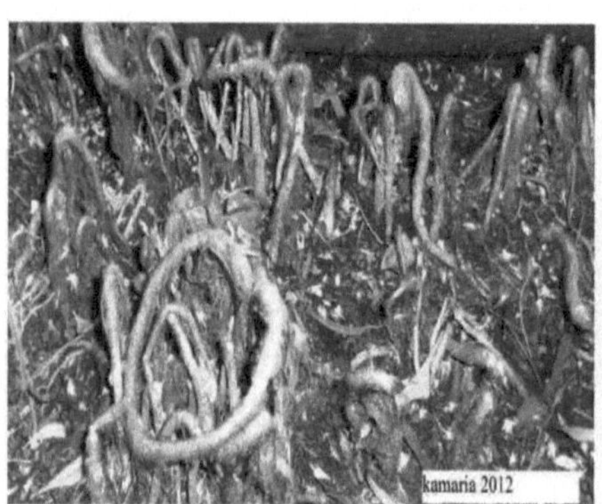

Photo 5 : Racines de *Heritieralittoralis*

Photo 6 : Contre-forts de *Xylocarpusmoluccensis*

Different types of roots in mangroves

Photo 7 : Chenal dans mangrove lagunaire à marée basse à Ikoni

Photo 8 : Peuplement de *Rhizophora mucronata* dans la mangrove lagunaire de Seleani

Photo 9 : Dégradation de la mangrove d'Ouellah par ensablement de front de mer

Photo 10 : Dégradation de la mangrove par coupe de bois

Photo 11 : Tanne vif de la mangrove littorale d'Ouroveni

Photo 12 : Individu d'*Avicennia marina* en dépérissement à voidjou

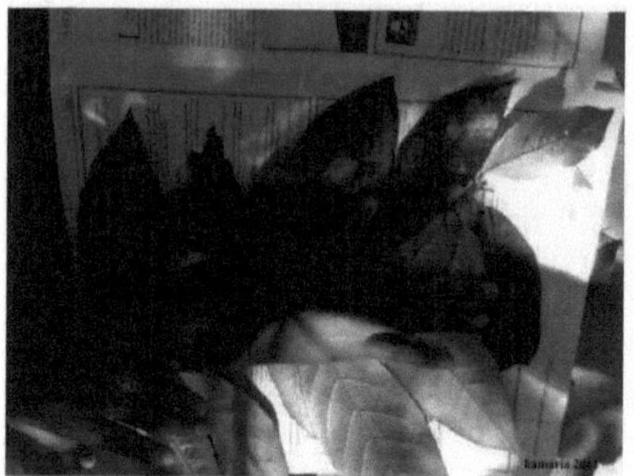

Photo 13 : Préparation d'un herbier

Photo 14 : Séchage d'un herbier dans une étuve

Photo 15 : Classification des spécimens sur les étagères

Photo 16 : Herbier marin

Mounting and conservation of herbaria

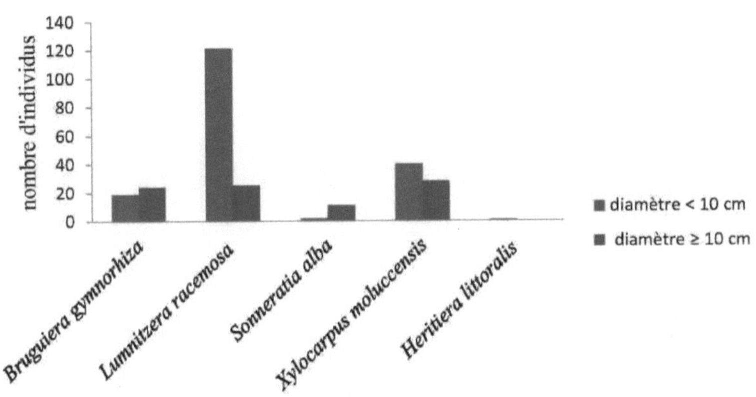

Figure 4: Graphical representation of the natural regeneration of mangroves in Ikoni

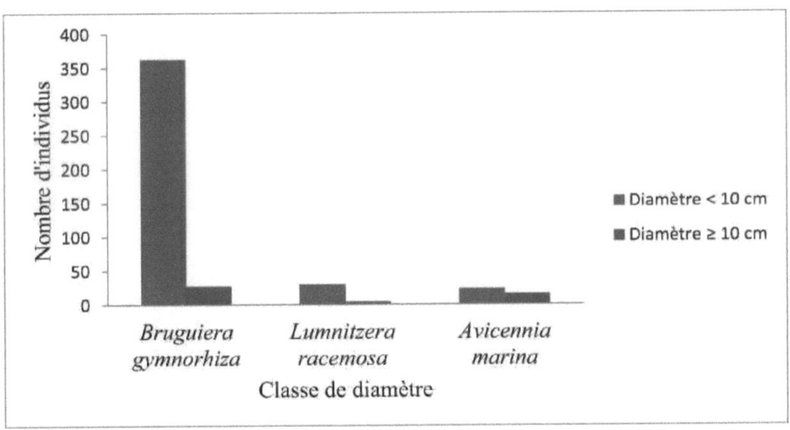

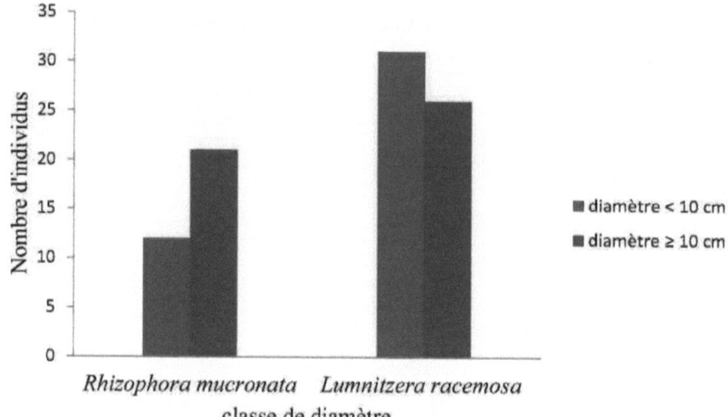

Figure 6: Graphical representation of natural mangrove regeneration at Seleani.

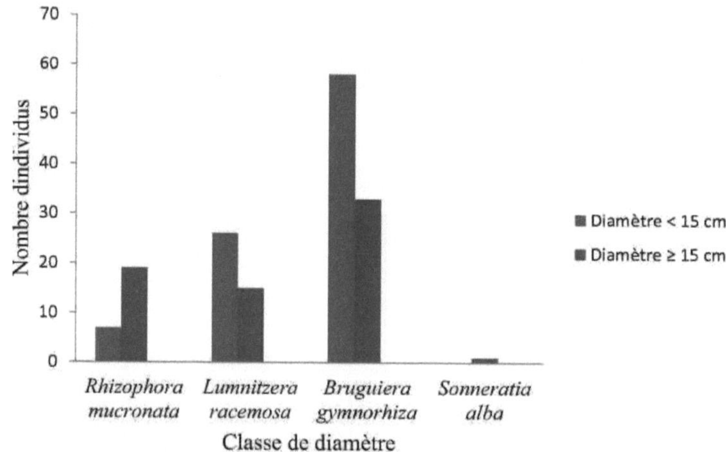

Figure 7: Graphical representation of the natural regeneration of mangroves in Domoimboini

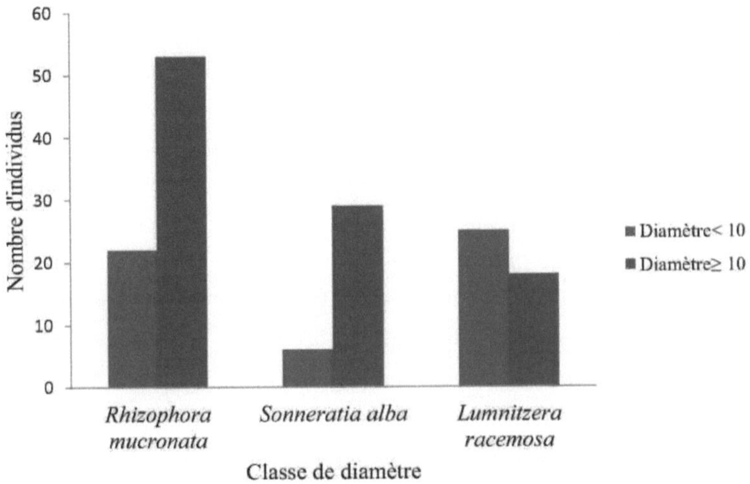

Figure 8: Graphical representation of the natural regeneration of mangroves in Ouroveni

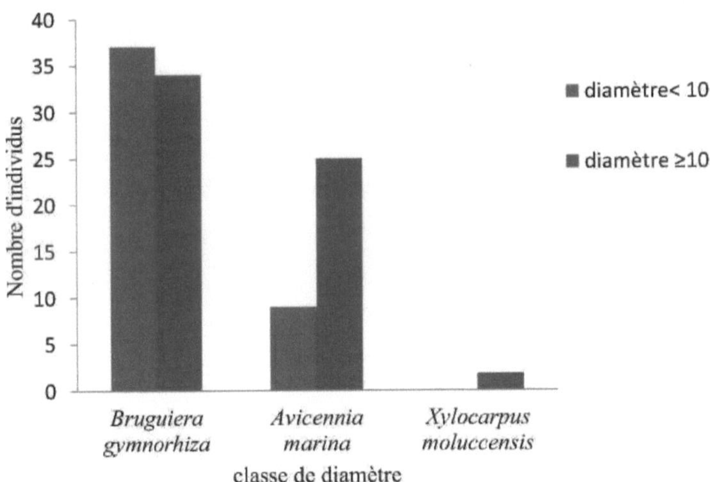

Figure 9: Graphical representation of natural mangrove regeneration in Ouellah

Summary

The ecological characterization of the mangroves of Ngazidja was carried out in order to provide a complete study concerning especially, the ecology of mangroves, their state of conservation, their mode of use, and their capacity to sequester carbon.

The study of these mangroves was carried out from biological inventories and ethnobotanical surveys. Indeed, seven species of mangroves *(Rhizophora mucronata, Avicennia marina, Sonneratia alba, Bruguiera gymnorrhiza, Lumnitzera racemosa, Xylocarpus moluccensis, Heritiera littoralis)* distributed in six families were described in this work. The study of natural regeneration shows an average rate of regeneration in some sites and a low rate in others.

Global warming exposes coastal areas to natural disasters. Faced with all these problems related to climate change, this work strives to find effective solutions to mitigate these phenomena. This study focused on the estimation of carbon stored by mangroves in Ngazidja. The results showed that carbon sequestration is proportional to the density and maturity of trees. Therefore, a good management of mangroves and a reforestation of mangroves can increase in a few years the quantity of sequestered carbon and consequently, reduce the rate of greenhouse gases (GHG).

The conservation and rehabilitation of mangroves is manageable. Their management depends on the participation of the State, the scientific community, the local population, in collaboration with experts and international organizations.

Keywords: Ecology, ethnobotany, mangrove, health status, threats, biomass, carbon stock, Comoros.

Framer: Doctor ROGER Edmond

Abstract

Ecological characterization of mangrove Ngazidja was conducted to provide a comprehensive survey mainly affecting the ecology of mangrove conservation status, how to use them, and their ability to sequester carbon.

The study of these mangroves was done from biological inventories and ethnobotanic surveys. Indeed, seven species of mangroves (*Rhizophora mucronata, Avicennia marina, Sonneratia alba, Bruguiera gymnorrhiza, Lumnitzera racemosa, Xylocarpus moluccensis, Heritiera littoralis*), in six families have been described in this work. The study of natural regeneration, regeneration shows an average rate at some sites and low in the other.

Global warming, coastal areas exposed to natural disasters. Faced with all these problems related to climate change, it is urgent to find effective solutions to mitigate these phenomenal. This study examined data on estimates of carbon stored in mangroves Ngazidja. The results showed that carbon sequestration is proportional to the density and mature trees. So, a good management of mangroves and mangrove reforestation can increase within a few years the amount of carbon sequestered and thus reduce the rate of greenhouse gas (GHG) emissions.

Conservation and rehabilitation of mangroves is controllable. Management depends on the participation of the State, the scientific community, the local population, in collaboration with experts and international organizations.

Keywords: ecology, ethnobotanic, mangrove, health status, threats, biomass, carbon stock, Comoros.

Advisors: Doctor ROGER Edmond